FOUNDATIONS FOR THE FUTURE: THE AICPA FROM 1980 TO 1995

STUDIES IN THE DEVELOPMENT OF ACCOUNTING THOUGHT
Gary John Previts and Robert J. Bricker, *Series Editors*
Weatherhead School of Management
Case Western Reserve University

FOUNDATIONS FOR THE FUTURE:
THE AICPA FROM 1980 TO 1995

by *Philip B. Chenok*
  AICPA President (1980–1995)
  with *Adam Snyder*

JAI PRESS INC
Stamford, Connecticut

Transferred to digital printing 2005

Contents

Acknowledgments

Many people contributed to this book. Some worked untiringly to support the Institute. Others reviewed and commented on the manuscript. In earlier years, still others gave me the foundation necessary to serve for fifteen years as president and CEO of the AICPA.

I am indebted to a current colleague at New York University, Professor Michael Schiff, who in earlier years taught me about accounting. During my years in public practice, I had the good fortune to be associated with a number of CPAs who helped me understand the profession and professionalism, including John C. Wilson, a partner in the New York City firm, Pogson, Peloubet, where I began my career, and Leroy Layton, Archie Mackay, Mort Solomon, Paul Clark, and John Thompson, my partners and friends at Main Lafrentz & Co.

At the Institute, I was privileged to work with many talented volunteers whose efforts shaped the changing nature of the profession and are the substance of the story told in this book. The chairs of the board of directors who served between 1980 and 1995, all of whom are listed in the Appendix, have my sincere appreciation for their commitment to the profession and to the Institute.

The AICPA staff is the glue that keeps the organization together. Several helped immeasurably: Don Schneeman, general counsel and secretary and my closest advisor; Tom Kelley, who headed our technical areas and whose untiring efforts made self-regulation work; B.Z. Lee, who after a full and rewarding career, during which he also served as board chair, helped reshape the legislative efforts in Washington, along with John Hunnicutt, who came aboard to assist and later succeed B.Z.; Ed Nemiec, who managed the business side of things; Geoff Pickard, who led the communications effort and who convinced me to take this project on and saw it to fruition; and Jo De Los Reyes, my assistant who made it all work. B.Z., Don, Tom, Geoff,

John, and, at a later stage, Ellen Goldstein, all added invaluable comments to various drafts of this manuscript.

The state societies are a most important partner to the Institute, and their staff leaders were critical to whatever successes we achieved between 1980 and 1995. Jack Brooks in Connecticut, Jim Kurtz in California, Mary Medley in Colorado, and Marty Rosenberg in Illinois all participated above and beyond the call.

Special thanks also to my collaborator, Adam Snyder, without whom this project would never have been started and certainly would never have been completed, and to former AICPA board member and accounting educator Gary Previts, who was a driving force in initiating the project and who served as a major cheerleader and commentator as it unfolded, reading countless drafts as the manuscript evolved. The AICPA library assisted Adam during his many visits there, and Marie Bareille shepherded the manuscript toward final publication. The Institution and the AICPA Foundation also financially supported this book, for which the

Finally, I am most grateful to the members of the Institute, particularly those who made the commitment to serve on committees and boards, for their participation in the profession and for helping to establish a "foundation for the future."

Foreword

The AICPA Foundation, which dates back to 1922, supports innovative programs to advance the accounting profession and, in particular, the quality and relevance of accounting education for professionals of today and tomorrow. As the AICPA's Vision Project carries out the vital job of looking toward the future, we at the Foundation are prepared to support programs to help achieve that vision. We also thought it was important to understand our past. Both self-evaluations are critical to our profession.

This is why we decided to partner with the AICPA in sponsoring *Foundations for the Future*, which documents the many significant events that impacted the accounting profession during Phil Chenok's term as AICPA president. This volume continues the path laid by the two books that came before it, authored by Phil's predecessors, John L. Carey and Wallace E. Olson.

The years covered in *Foundations for the Future*, 1980–1995, were critical ones for our profession. As AICPA membership doubled, the nature of services that we as a profession offered our clients was dramatically transformed. The complexity of business transactions and reporting obligations increased exponentially. These drew both public and private sector attention, and the profession again stepped forward to shape timely responses.

Indeed, the profession was on the cutting edge of the dramatic changes that occurred within the U.S. business community during the fifteen years of Phil's tenure. More than ever before, it had a critical role in helping to ensure our nation's financial well-being. *Foundations for the Future* tells the story of how the profession adapted to these changes and the challenges that accompanied them.

The AICPA Foundation has always been particularly concerned with helping to broaden the profession. Many of our activities, for example, are aimed at providing students from different racial and ethnic backgrounds greater access to accounting education. It is also why we wanted *Foundations for the Future* to be written in a way that would appeal to as diverse an audience as possible.

The AICPA Foundation's ultimate educational goal is to help attract talented college graduates and to retain them in the profession. With future accountants in mind, it is our hope that *Foundations for the Future* not only lays down the facts for future generations, but also portrays our profession as an important, exciting place in which to work.

Dominic Tarantino
Board of Trustees President, AICPA Foundation

Introduction

The fifteen years between 1980 and 1995 were a period of tremendous change in U.S. business, and therefore for the accounting profession. The computer revolution, the transformation from an industrial-based economy to a service oriented one, the globalization of business, and the expansion of legal liability were only a few of the major shifts that occurred during my watch as president of the American Institute of Certified Public Accountants (AICPA), the national professional membership organization for CPAs.

It is no wonder that AICPA membership was transformed as well. Total membership almost doubled between 1980 and 1995, from about 161,000 to almost 325,000. Of equal significance, those working in business and industry jumped from 35.5 percent of total membership to 41.7 percent, and for the first time outnumbered those working in public accounting firms.

The reason for the shift in our membership was that accountants broadened their job descriptions because clients and employers were demanding an ever-increasing array of services. For those in public practice, the audit remained a vital but limited discipline, so in order to grow and remain relevant accounting firms had to diversify. Indeed, by 1995 CPAs could be found throughout the ranks of every conceivable type of business, from law to entertainment to manufacturing to health care, and in almost every business capacity. They organized a company's financial records, and prepared tax returns. As chief financial officers, they were part of top management responsible for setting prices, establishing research and development budgets, and plotting competitive strategies. I don't think it's an exaggeration to say that within many U.S. businesses, often it is a CPA who makes the difference between profit and loss, between success and failure.

But there is an irony involved in the way accountants are perceived. On the one hand, we are often stereotyped as uncreative number crunchers, akin to Scrooge's bespectacled bookkeeper Bob Cratchit, interested only in making sure the numbers add up. On the other hand, in today's complex business world we are increasingly asked to be among the most creative of businesspeople. Taxpayers want their advisors to devise the most imaginative strategies possible under the law. Corporations want their accountants to create innovative systems that will allow their businesses to operate at maximum efficiency. And investors expect auditors to uncover the most subtle secrets within a company's financial statements to help them evaluate current performance and future prospects.

THE CPA

A CPA is trained and licensed to conduct "assurance" services, namely, services that "assure" other people of the accuracy and validity of information. As independent auditors, CPAs play a unique role in our society, different in a way from any other profession. We're fundamentally a group of professionals assessing information communicated by organizations to others who, in turn, use that information to make decisions. The information may be used by lenders to gauge the advisability of making a loan, investors to predict a company's financial future, or governments to assess taxes or determine whether certain regulatory reporting obligations have been met. Except for the *Good Housekeeping* seal of approval, what is another example in our society of a professional group in effect assuring the communication of someone else's information?

THE AUDIT

In the United States, the genesis of today's financial statement audit can be traced to the eighteeth century, when accountants were employed by English property owners to validate the financial information they were receiving from the colonies. Since then, the accountant's functions evolved through a number of dramatic transformations, but the audit remained the fundamental reason for licensing CPAs.

An "audit," as defined by *Webster's Seventh New Collegiate Dictionary*, is "a methodical examination and review" or, more specifically, "the final report of an examination of books of account by auditors." Among the many different functions provided by today's CPA, the audit remains the most prominent. That's because the Securities Act of 1933 and the Securities Exchange Act of 1934, passed by Congress in response to the vast sums lost by investors in the stock market crash of 1929 and the subsequent financial depression, require all public companies in the United States to be audited every year by an independent auditor. The federal government subsequently expanded these requirements so that pension plans and local and state governments also had to be audited. Audits also became part of many real estate and other commercial agreements, as well as bank loan requirements.

THE FINANCIAL STATEMENT

If accounting is called the language of business, the financial statement is its communications device. Financial statements tell a story; a story about how well or poorly a company has fared over a period of time and where it stands today. In large part, the financial statement tells this story by disclosing income and expenses and what the company owns and owes. Like any communications device, financial statements have a set of common ground rules understood by both those who tell the story and those who read and interpret it. These ground rules, which were developed over a period of many years, are called generally accepted accounting principles (GAAP).

The AICPA

Most CPAs in the United States belong to the AICPA, the national professional organization responsible for establishing auditing standards and for enforcing a code of professional ethics and a high level of quality on all CPA practices. At one time, the AICPA also set all accounting standards, but, as will be discussed later, that function primarily now rests in the hands of the independent Financial Accounting Standards Board (FASB). As a membership organization, the AICPA strives to improve the lot of its members, but it also operates under the concept that one of its most important roles is to help its members working in public accounting firms serve the public interest.

One of the distinguishing characteristics of a profession is a set of ethical standards designed to motivate honorable behavior. Prior to 1988, the profession's code of ethics was a little like the Ten Commandments in that they decreed what "thou shall not" do, rather than what "thou shall" do. After the AICPA membership voted to frame the code in a more positive way (see Chapter Four), the quality of practice improved significantly.

At the foundation of our rules is that a CPA must be independent, meaning that the auditor has no financial interest or affiliation with the entity being audited. He or she must also be licensed. In 1896, New York became the first state to enact a law providing for the licensing of CPAs and giving them the exclusive right to express an opinion, as a CPA, concerning the conformity of financial statements with existing standards. Since then, all fifty-four U.S. states and territories have established licensing requirements. While these requirements continue to differ among the jurisdictions, we made great strides between 1980 and 1995 toward adopting some degree of uniformity as far as the "three Es"—education, examination, and experience—were concerned.

During this period we also significantly strengthened the requirements of AICPA membership. Because standards are a moving target, continually changing with the ever-increasing demands of financial trends and legislation, the AICPA determined that a CPA's education requirements should not end with passage of the uniform CPA examination, which all CPAs in all jurisdictions must pass in order to become licensed. (See Chapter Five.)

Since 1989, the AICPA has also mandated continuing professional education (CPE) requirements. In order to retain membership in the Institute, AICPA members with an accounting or auditing practice must have 120 hours of CPE during every three-year period, with a minimum of 20 hours per year. Members in business, industry, and education must have 60 hours, with a minimum of 10 hours per year. Moreover, all AICPA members in public practice must undergo a periodic review by other CPAs of the quality of his or her accounting or auditing work.

The decision in the early 1980s to impose some fairly stringent AICPA membership requirements, including CPE and peer review, carried with it the significant risk that many CPAs would simply choose to revoke their membership in the Institute. After all, there is no obligation on the part of CPAs to belong to the AICPA. But to the credit of the AICPA and its members, not only were these requirements overwhelmingly approved, but membership continued to increase. There may have been a little self-interest involved, of course. Members in public practice recognized that shoddy work could make them vulnerable to lawsuits which could be much more expensive than either CPE or the peer review process. In fact, legal liability, as we shall see in Chapter Two, became a tremendous concern to the profession, costing the largest five accounting firms (the Big Five) alone more than $1 billion per year and mushrooming the cost of professional malpractice insurance for all CPAs.

Chapter 1

Strategic Planning: The Times, They Were a Changing

When I became AICPA president in July 1980, I was immediately bombarded with a host of issues that threatened to revolutionize the way CPAs would conduct themselves in the coming years. All the rapidly moving trends we were seeing in society and in business, everything from the computer revolution, to globalization, to the increase in class action suits, were clearly having a revolutionary effect on the accounting profession. Trying to come to grips with these issues, particularly how to manage them in a context beneficial to CPAs, seemed an extremely daunting task as I took over the reins of an organization with more than 160,000 members and an annual budget of $40 million.

I had my work cut out for me, but was fortunate during these early months to have the benefit of the guidance of my predecessor, Wally Olson. It wasn't until two years later, however, when he published his own account of his AICPA presidency, that it became apparent Wally had had similar thoughts during his tenure. One of the main theses of his book, *The Accounting Profession: Years of Trial*, was that the profession had been caught largely unprepared to react to the social and political trends of the 1960s which came to full flower in the 1970s. As a result, Wally believed that the measures initiated between 1969 and 1980 represented a belated effort by the profession to respond to these trends. I took that to heart as I immersed myself in the issues at hand.

In order not to be overwhelmed by events outside our control, the AICPA needed a mechanism to help clarify "moving target" issues that threatened to so profoundly change the way we lived and worked. It also occurred to me, as I reflected on the nature of these events and how they had already caused concern and disruption among our membership, that with a little foresight perhaps many of them could be anticipated earlier in the process. By anticipating events, I hoped we could in some way guide them, or at least be better prepared for them as they arrived.

Fortunately, about a year after my appointment, George Anderson, who later would chair the seminal AICPA Committee on Standards of Professional Conduct (see Chapter Four), became chairman of the AICPA board of directors. George shared my conviction of the importance of looking beyond today and into the future.

FUTURE ISSUES COMMITTEE

Both George and I believed that the Institute could be more proactive in identifying and acting on issues affecting the profession. Together we decided to appoint a Future Issues Committee charged with developing a list of issues that the Institute ought to be focusing on today, in order to forestall problems tomorrow. The committee, in other words, became a device for monitoring the most important issues facing our profession and directing those issues to the appropriate committees for further study and review. It was chaired by Richard Hickock, a partner in Hurdman Cranston. Its other members included representatives from accounting firms (Bruce Dixon of Ernst & Whinney, Arthur Wyatt of Arthur Andersen & Co., and R. Scott Elsasser of Seidman & Seidman), educators (Ernest Davenport of Howard University and Charles Horngren of Stanford University), corporate accountants (Paula Cholmondeley of Westinghouse Elevator and Robert Thorne of the U. S. Gypsum Company), and a state society representative (Jack Brooks, executive director of the Connecticut Society of CPAs). As a de facto member, I made it a point to try to attend every meeting.

After an initial session in July 1982, the members of the Future Issues Committee met quarterly for the next two years. We began by consulting with John L. Carey, who as head of the Institute for many years until the late 1960s had helped it establish some long-range objectives. He was also the author of *The Accounting Profession—Where Is It Headed* (1962) and *The CPA Plans for the Future* (1965), as well as the two-volume work, *The Rise of the Accounting Profession* (Volume 1, 1969; Volume 2, 1970), the prototype for this text. He, therefore, had an excellent perspective on the profession's history and the issue of future planning.

The committee members reviewed dozens of publications and research studies relevant to the topics that we began to identify, and consulted with various futurists who helped us develop techniques we could use for exploring the future. One of our consultants, for example, was Dr. Burt Nanus, professor of management and director of the Center for Futures Research at the University of Southern California Graduate School of Business Administration.

The central focus of our inquiry was to identify the issues most likely to significantly impact the accounting profession. One early problem was that we soon realized we were finding far more issues than we could manageably study. On our first pass we identified 134. But by merging and prioritizing them, we were able to come up with the following far more manageable list:

1. Expansion of services and products offered by CPAs and accounting firms
2. Changes in the nature and extent of competition
3. Widespread computerization and automation of business operations
4. Litigation and legal liability as an increasing problem for the profession
5. Increased specialization among accountants
6. Accounting standards overload
7. The role of self-regulation
8. Upward mobility of women
9. Improving the quality of practice
10. Major reform of the federal income tax system
11. Changes in the composition of AICPA membership
12. Maintaining independence and objectivity
13. Diversity in CPA qualification and performance requirements
14. Mission, goals, and objectives of the AICPA

It wasn't yet our task to suggest what the profession should do about these fourteen issues. That came later. We did, however, make a few specific recommendations that ultimately brought us to that goal.

In the first place, we recommended that the Institute undertake a comprehensive analysis of each issue. But more than that, we wanted the Institute to adopt an overall commitment to become more future oriented. This included giving the Future Issues Committee an ongoing role, implementing a trend-monitoring system to provide early indications of future issues, and, most important, establishing a formal strategic planning process to make assumptions about the future and to develop goals, objectives, and strategies that would turn those assumptions to our advantage.

For most of the fourteen issues, we either referred them to committees already working in those areas, or established new committees to deal with them. During these years, for example, the Institute's board of directors authorized the formation of a special committee to study the upward mobility of women, an accounting standards overload task force, and a technology subcommittee of the Institute's management advisory services (MAS) division. The special committee on financial reporting (known as the Jenkins committee), formed in 1991 to address concerns about the relevance and usefulness of business reporting, also had its genesis with our initial 1984 report of the Future Issues Committee. (The Jenkins committee is discussed in detail in Chapter Six.)

In most cases, however, we made available our findings to existing committees and asked them to devise a program that would address our concerns. In October 1983, for example, the Special Committee on Standards of Professional Conduct was just getting started, and we asked them to include in their deliberations three issues we had identified as particularly important: expansion of services and products, changes in the nature and extent of competition, and independence and objectivity.

A MISSION STATEMENT

In some cases, we took it on ourselves to take action. One such instance concerned what was perhaps our most important recommendation—the need to spell out an AICPA mission statement. We felt strongly that the Institute needed to clearly state the raison d'être of our 100-year-old organization. Only then could we develop a plan to move the organization in the direction of meeting our stated objectives. In other words, before we could collectively move toward a common goal, we had to agree on the nature of that goal.

Another reason we were so eager to develop a mission statement was that we knew it would provoke a confrontation between two opposing points of view, and hoped it would settle the issue once and for all. In one camp were those who believed CPAs operated as a public trust, and that the profession's first responsibility was to protect the public good. Those in the other camp believed that as a membership organization, the AICPA's goals and activities should primarily be directed toward the benefit of its members. They believed we should be spending our resources to promote the indispensability and professionalism of CPAs and to service our constituency through such activities as a comprehensive library and help line, extensive CPE curriculum, and an effective Washington office that looked after our legislative interests.

Also during these years, the Institute developed programs to provide small firms with services similar to those available within large practices. Our Technical Information Service (TIS), for example, allows any AICPA member to ask for assistance on any accounting or auditing problem. The most frequently asked questions and answers are also published in AICPA Technical Practice Aids.

All these activities are vitally important, of course, and are all significant responsibilities of the Institute to its members. But I also felt very strongly that our organization's primary purpose was to function as a public interest organization, and that it was necessary to maintain that integrity. The debate about whether we are primarily a professional association (setting standards and guarding our professionalism) or basically a trade association (product and service support and member advocacy) continues to this day.

I always saw CPAs as members of a profession guarding the public interest, and believed that also should be the primary goal of the Institute. But what is the public interest? The AICPA's Code of Professional Conduct defines it as "the collective well-being of the community of people and institutions that the profession serves." In fact, concern for the public interest is part of our professional designation—certified *public* accountants. As certified *public* accountants, we have always been charged with a public interest responsibility, and over the years the public has come to rely on us in a wide range of areas. Most important, we have been given a "franchise," an exclusive right to express opinions on financial statements. This franchise was given because of the public's concern about the potential for harm if individuals or organizations acted on misleading information. Congressman John Dingell, during the congressional hearings he chaired in the 1980s (see Chapter Three), was correct when he said that our economic system works on trust. Those who use financial information must trust that the information they use is relevant and reliable. That's where CPAs come into the picture.

Shortly after the first Future Issues Committee report was issued, another committee was appointed to develop a mission statement for the Institute. It was headed by educator Sandy Burton, a former chief accountant of the

SEC. The committee drafted a mission statement, approved by the AICPA board of directors and then ratified by the governing Council, a 250- to 300-member body that meets twice a year to vote on changes in the AICPA bylaws and ethical standards. The Council has the authority to establish the policies and procedures of the Institute and to enact resolutions binding on the AICPA's board of directors and its officers, committees, and staff. Every state (as well as the District of Columbia, Guam, Puerto Rico, and the Virgin Islands) has a delegate designated by its state society, with additional delegates apportioned according to the state's size. There are also twenty-one at large Council members nominated by the board of directors and confirmed by the full Council. Board members and all past AICPA chairs and presidents are also members.

The mission statement as ratified by Council read:

> The American Institute of Certified Public Accountants is the national pro-fessional organization for all Certified Public Accountants. The mission of the AICPA *is to act on behalf of its members* and provide necessary support *to ensure that CPAs serve the public interest* in performing *the highest quality professional services.* In fulfilling its mission, the AICPA gives priority to those areas where public reliance on CPA skills is most significant. (Emphasis added)

That was our mission statement during my tenure as AICPA president. In November 1995, the strategic planning committee proposed a few changes in the statement, which the board of directors subsequently ratified. The new mission statement now reads:

> The American Institute of Certified Public Accountants is the national profes-sional organization for all Certified Public Accountants. Its mission is to provide members with the *resources, information and leadership* that enable them to provide *valuable services* in the highest professional manner to benefit the public *as well as employers and clients.* (Emphasis added)
>
> In fulfilling its mission, the AICPA works with state CPA organizations and gives priority to those areas where public reliance on CPA skills is most significant.

A close reading of the change suggests that the "public interest" versus "members interest" issue is still an open question.

A FORMAL STRATEGIC PLANNING PROCESS

In addition to helping to identify future issues and instituting a mission statement, the other important contribution by the Future Issues Committee was its call for the establishment of a formal strategic planning process. This resulted in the formation of a Strategic Planning Committee, which I chaired throughout my tenure as president. Its initial membership included two other people who also served on the Future Issues Committee—Paula Cholmondeley from industry and Marvin Strait, a CPA with his own small practice in Colorado Springs who later became chairman of the AICPA's board of directors. Both Marvin and Paula, a top executive at Westinghouse for many years and an expert in long-range planning, were tremendously helpful during these years.

In our first report, published in October 1988 and entitled *Strategic Thrusts for the Future*, we began by asking ourselves, "What is the future going to look like for the accounting profession?" In trying to find answers, we soon found it helpful to divide the committee into three groups. One was charged with making assumptions about the future of individual CPAs, another about the future of accounting firms, and the third about the future of the AICPA as a professional organization. Each group developed a list of its own assumptions, complete with a description of the issue, its background, and what was being done currently, if anything, to influence the issue.

Next we merged the three lists. In some cases the same assumption was made by all three groups. In other cases assumptions made by the three groups were contradictory, so we had to come to some common ground before issuing recommendations for action. With each assumption, we went through the process of asking ourselves: If we are correct, will the consequences be positive or negative? If we expected a positive outcome, we asked ourselves how we could ensure it occurred. If we believed it would have negative repercussions, we asked ourselves what we could do to help prevent them from occurring.

The Strategic Planning Committee during these years developed twelve overall strategic thrusts, backed by ninety specific strategic planning assumptions representing our view of the future. Five of the twelve thrusts—computerization, legal liability, specialization, improving the quality of the CPA practice, and the changing demographics of the Institute's membership—coincided with those issues identified by the future issues report four years earlier. This was an ongoing and evolving process, however. Subsequent Strategic Planning Committee reports, issued in 1990, 1991, 1993,

1994, and 1995, presented their own assumptions and action plans, which changed as the business environment changed. Sometimes assumptions or recommendations issued in one report were updated in the next report. Other times assumptions were reversed, or dropped altogether or until subsequent events revived the issue.

The important thing was that now we had a process. Events no longer controlled our actions. To a certain extent at least, we now had the ability to anticipate the events that were so profoundly affecting our profession and to take preemptive actions to turn them to our advantage.

I'm pleased to report that the strategic planning process begun in 1982 continues today with the even more ambitious CPA Vision Project. Its stated goal is to focus on the future today so the CPA profession can seize the opportunities of tomorrow. By spending considerable resources to reach out to thousands of CPAs across the country in hundreds of focus groups (called future forums), this ongoing visioning process should allow CPAs continually to plan for the future.

"We realize that we owe a debt of gratitude to the future issues and strategic planning processes that came before us," says Jeannie Patton, the executive director of the Utah Association of CPAs who served as director of the Vision Project. "The process is going to ensure that CPAs continue to focus on and better anticipate the future, as well as address the inevitable challenges and opportunities. By identifying the issues that our grassroots membership tell us will be most meaningful to the profession, we can be proactive, not just reactionary." (Sound familiar?)

DIVERSIFICATION AND SPECIALIZATION

It is interesting that the fourteen issues identified in the initial 1982 future issues report could almost serve as a table of contents to this book. Litigation and legal liability, for example, will be discussed in detail in Chapter Two, self-regulation in Chapter Four, standards overload in Chapter Six, the Internal Revenue Service in Chapter Three, and changes in AICPA membership, including the upward mobility of women, in Chapter Seven. But let's for the moment take a snapshot of a few of these issues, beginning with the diversification and specialization of accounting practices, which permeate virtually every aspect of the profession and therefore don't fit neatly into chapter categories.

The functions most associated with accountants are auditing and tax preparation, yet the typical CPA increasingly did much more than that during the years between 1980 and 1995. In fact, the more CPAs learned about a particular industry, the more their clients or employers began to expect an expanded array of services. This is one of the reasons why consulting became the fastest growing segment within the profession, and why the largest accounting firms in the United States began to take in less than half their revenue from attest services.

That doesn't mean the audit didn't continue to be the cornerstone of our profession. In fact, the complexity and significance of auditing the largest multinational corporations should not be underestimated. Auditing the likes of a General Electric, an IBM, or a Ford Motor Company takes hundreds of CPAs working full time for months, and requires sophisticated financial and management expertise.

But having said that, there is also a limit to the amount of audit work available. There is a finite number of public companies at any one time, so there is a finite amount of revenue to be derived from those audits. But there is virtually no limit to the myriad consulting services that can be offered by accounting firms. Some people have even called the audit a "loss leader"—a willingness to lose money for the opportunity of examining the intimate workings of a large corporation in the hopes of identifying the kinds of prospective consulting services that might benefit the company the most.

But the move into consulting was driven by much more than simply a desire to grow. There was a seminal change in accounting, which made diversification into other areas necessary for our very survival. Probably the single most important factor driving this change was the personal computer (PC), which in less than two decades revolutionized the way businesses operate, keep their books, pay their taxes, and access information. It certainly affected audits, as we predicted in our 1984 future issues report when we wrote, "Advances in the audit process resulting from the increased use of computers may allow firms to improve audit efficiency and to audit on a continuous basis."

In hindsight, no prediction of the effect computerization would have on business ever proved to be underestimated. But PCs had a particularly profound effect on the accounting profession. That's mostly because accounting firms have traditionally relied on time-based revenue. Clients pay a fee according to hourly rates for specific functions—an audit, tax return preparation, and so on. But the computer made that traditional fee model untenable.

In the first place, the PC allowed accountants to perform detailed functions, such as tax return preparation, so quickly that charging by the hour for these services was no longer a viable business strategy. The model had to be changed so that clients began paying for knowledge, not time. Second, an increasing amount of work that historically had been performed by accounting firms—particularly smaller firms—could now be done by the clients themselves. Just a few years ago, an accountant would come into a business and spend days poring over a company's records in order to "write up" the books. That may seem like an image from the nineteenth century, but in fact it was the norm as recently as twenty years ago. Yet by 1990, with a $3000 investment for the latest PC and a $70 piece of software such as Quickbooks, small businesses could track their inventory and accounts receivable, send out invoices, pay their bills, balance their checkbooks, and prepare financial statements and tax returns all by themselves.

"Can the profession," we asked in the 1984 future issues report, "adapt to take advantage of the widespread computerization and automation of business, including the widespread use of personal computers?" We predicted that within the next two decades the increasingly rapid growth in computer and information technology would revolutionize business practice.

Well, it didn't take two decades, and given these new realities it's little wonder that accounting firms of all sizes had to branch out into new areas just to survive. Because of their training and experience, there was virtually no limit to the consulting services CPAs were asked to provide, including systems design, computer installation and programming, strategic planning, global marketing, and a variety of different ancillary services. Accountants became executive headhunters, tax specialists, lobbyists, expert witnesses, financial planners, actuaries, and financial engineers. And they do everything from planning urban shopping centers, to testifying in multi-billion dollar lawsuits, to keeping track of scoring at the Olympic games, to designing Third World transportation infrastructures. Many CPAs began to specialize in evaluating the worth of businesses in preparation of a sale, inheritance, or stock offering. Another growing specialty was litigation support, in which CPAs help law firms maintain an indexed record of who said what about whom in preparation for trial.

ACCOUNTANTS VERSUS CONSULTANTS

Within some firms, the expansion into consulting created a clash between two cultures, igniting a firestorm that has yet to be quelled. Consultants were pitted against the traditional auditor, who tended to be more professionally remote (independent) than their new, more aggressive, sales-oriented consulting colleagues.

While in 1995 traditional accounting services still constituted a majority of total revenue at most of the large accounting firms, this eventually changed, as the consulting side began to experience annual growth of 20 percent or more. It has probably now reached the point where the line between a consulting firm and an accounting firm is completely blurred.

The dispute was more than cultural, of course. As with most business disputes, in the end it came down to money. Nowhere was the conflict between accountants and consultants more acrimonious than at Andersen Worldwide, which in 1989 spun off its consulting business into a separate unit called Andersen Consulting. In 1996 Andersen Consulting accounted for $5.3 billion of Andersen Worldwide's total revenue of $9.5 billion. According to the *Wall Street Journal*, each partner in Andersen Consulting generated nearly $1 million in profit, compared with $600,000 per accounting partner. This had the consulting side clamoring for more influence in decisions that affected the entire firm, including, of course, profit share participation.

COMPETITION

Diversification cut both ways. On the one hand, it gave CPAs the opportunity to branch out from traditional attest services. But on the other hand, non-CPAs entered areas that traditionally were the exclusive domain of CPAs.

The single event that best illustrated the encroachment of non-CPAs into what were historically CPA functions—and which most alarmed many small to midsize practitioners—was the decision in 1991 by the American Express Company to begin acquiring accounting firms and incorporating them into a new division called American Express Tax and Business Services. The financial giant eventually acquired almost 100 smaller accounting firms. Because of questions about whether firms owned by non-CPAs should perform audits, American Express left the auditing functions of its acquired accounting firms with the CPA owners. It took only the remaining accounting and tax services for itself.

The initial reaction was to try to block the acquisitions by American Express, or at least restrict CPA employees from informing the public of their CPA designation unless they worked for a firm that was 100 percent owned by CPAs, which of course American Express was not. But in 1995 a federal court in Tallahassee effectively put a stop to that strategy when it ruled that the Florida board of accountancy could not legally prohibit CPAs working for American Express in Florida from advertising their credentials. And although some states still require a CPA firm to be 100 percent owned by CPAs, in 1995 the AICPA adopted a rule allowing some non-CPA ownership of CPA firms. Most states have subsequently followed suit.

When American Express first began acquiring accounting firms, there was a great cry among many members of the profession that this was blasphemy, that American Express Tax and Business Services competed unfairly with local practitioners. But I always felt these fears were misplaced. There's nothing fundamentally wrong with someone in the public marketplace deciding to finance an accounting service organization. The issue is not so much about ownership, but rather quality control.

As long as the independent auditor adheres to professional standards in delivering services, it doesn't make a difference who owns the organization. It could be American Express, or John Doe, CPA. American Express was simply attempting to use CPAs to satisfy a perceived market need, just as the local CPA firm must do. After all, American Express was acquiring accounting firms to achieve its goals. CPA firms ought to be smart enough to see this as an opportunity rather than a threat. As the world changed and became more complex, there were ever increasing needs for professional help, including retirement and estate planning, global positioning, and other complex financial transactions. CPAs have to shed themselves of old ways and take a fresh look at marketplace needs. Every successful professional should see change as an opportunity. Those who don't will just eke out a living, barely getting by, which should be anathema to anyone entering or seeking success in the accounting profession.

The issue is similar to the one facing the medical profession. The criticism of health maintenance organizations (HMOs) is not that doctors are employed by a commercially operated, for-profit organization. Consumers are concerned that their physicians provide them with quality care, not who signs the doctor's check. The system breaks down only when owners seek to influence the quality of service, something some HMOs have clearly tried to do.

That's why the accounting profession must be so diligent about establishing and enforcing a comprehensive set of professional standards. Regardless of who owns a CPA firm, it is the CPAs themselves—monitored by their states and the AICPA—who are ultimately responsible for complying with those standards.

ACCREDITATION OF SPECIALTIES

It should have been no surprise to anyone in the profession that we were going to experience an ever-increasing wave of competition from non-CPAs. If it hadn't been American Express, it would have been another entity, so even without American Express entering the picture competition would not have slowed. Indeed, we recognized that inevitability in our 1984 future issues report. Given what we knew was going to be a dramatic change in the nature of U.S. accounting firms, how, we asked, should the profession adapt so that firms could take maximum advantage of opportunities to expand services and products in a manner appropriate to the professionalism and integrity of CPAs? And recognizing that nontraditional services would fuel the growth of the profession, should we establish a formal system for recognizing and accrediting these new specialties? If so, we would have to recognize that they may be far afield from traditional attest services such as auditing and reviewing financial statements.

This issue of formal recognition of specialties had been extensively discussed and debated during the decade before my appointment as AICPA president. A special committee on the scope of professional services and specialization had been appointed, but generated little tangible results. In 1978, that committee concluded, "there is both a public need and a need on the part of the profession for a program of accrediting CPA specialties. Perhaps the most compelling evidence of that need is the widespread de facto specialization that already exists in the profession, presently based only on self-declaration by the individual or his firm."

Nevertheless, in 1980 the interpretation of rule 502 of the code still read, "Claiming to be an expert or specialist is prohibited, because an AICPA Program with methods of recognizing competence in specialized fields has not been developed, and self designation would be likely to cause misunderstanding or deception."

Finally, in September 1981, as our prohibition on self-designation of specialties was coming under attack by the Justice Department, our ethics prohibition on designating specialties was withdrawn. Thereafter, any complaint regarding self-designation was dealt with only on the basis of whether the statements made were false, misleading, or deceptive.

Simultaneously, however, there was a growing sentiment within the profession that the Institute should accredit certain specific specialties. In particular, the Colorado state society urged the Institute to move forward with such a program, and when the Institute failed to take timely action Colorado initiated a program of its own. At about this same time, the

California state society was establishing a similar program. And to the dismay of many in the profession who felt that CPAs were already sufficiently regulated, other states and organizations joined the bandwagon by urging federal or state licensing of personal financial planners.

Ultimately, in the spring of 1986, the Institute established a special committee on specialization to develop a program to recognize specialties through a formal accreditation process. This was simply a recognition of the inevitable. As more and more new business disciplines emerged, accountants were in fact specializing in them, no matter what state or federal agencies, or even the AICPA, had to say about them. The dispute centered on whether these specializations should be accredited by the profession.

The argument in favor of accreditation was that by providing CPAs with a professional recognition of their expertise, they would be in a better position to compete with non-CPAs. CPAs working in small and midsize accounting firms, however, who have always made up the lion's share of the Institute's membership, generally opposed additional accreditations. Many of them were concerned about overregulation, and had no desire to have additional examinations or other requirements thrust on them. They also feared that with just a few partners, they would have a difficult time accrediting even one partner in any particular area. Early certification proposals, for example, required CPAs to spend as many as 750 hours per year in a particular area of expertise in order to become accredited. Although large firms easily could dedicate one or more partners to any specialty, this was often impossible for smaller firms. On the other hand, the large firms weren't necessarily keen about new accreditations either because they were for all intents and purposes already accrediting people within their firms, just by presenting them to clients as specialists.

Some members also feared that specialty designations would pit CPA against CPA by giving an advantage to someone with an accreditation. For example, expert witnesses without a designation of expertise might over time stop being called to testify. Or if they were called, their testimony might not seem as credible.

These points of view were understandable, but the last thing we wanted to do was to repeat the mistakes of the American Medical Association (AMA). As the classic family doctor gave way to an increasing number of medical specialties, general practitioners nevertheless continued to control the inner workings of the AMA and prevented it from recognizing or accrediting specific specialties. As a result, specialists created their own organizations, so groups like the American Academy of Pediatrics, the

American College of Surgeons, and the American Academy of Orthopedic Surgeons began growing at a significantly faster clip than the AMA. In fact, AMA membership has actually fallen in recent years, and in 1997 its 292,628 members represented less than half of all physicians in the United States.

Nevertheless, accreditation of CPAs did not receive much support during most of my tenure as AICPA president. In 1986, for example, the Institute's governing Council approved the development of a process for accreditation of specialties, but by 1989 only personal financial specialists had been recognized. There was, however, a groundswell of support for this one specialty, which used to be called estate planning and typically was targeted at only wealthy individuals. But as Americans began to live longer, invest more on their own, and retire earlier, they often needed help from a professional. They also needed assurances that the person advising them was qualified and met certain ethical standards such as objectivity, honesty, and competence.

For similar reasons, in 1992 an accredited specialty in business valuation service was approved, although it wasn't finalized until 1996. Establishing the worth of a company is often critical for the purposes of buying or selling a business, and for estate planning. We also developed new membership divisions for such specialties as tax, management advisory services (MAS), and information technology.

Right to the end of my AICPA tenure, however, accreditation continued to be a controversial issue. At the May 1994 Council meeting, Ellen J. Feaver, chairman of the Special Task Force on Accreditation, presented the committee's final report. It recommended specific guidelines for any official AICPA accreditation, including a special examination, 450 hours of experience, and 60 hours of CPE every three years. Any new credential would also have to demonstrate clear evidence of a demand among members, including a commitment on the part of members to take the first examination.

The report was approved by the Council, but not unanimously. Phil Doty, a Council member from Colorado, urged its rejection because "while many of the seven individual recommendations have merit, the cumulative effect of satisfying all of them will make it very difficult to create new credentials."

Donald Dale, president of the Virginia CPA Society, opposed the number of hours of experience and CPE required to maintain the specialty. "As specialization comes on line for business valuation, computer consultant, contractor auditing, nursing home advisor, employee benefit plan specialist, and so on, most of us here today are not going to be able to qualify. And as insurance companies, underwriters, and bankers begin to mandate that without a particular specialty designation we are not qualified to perform a particular type of work, this will have a disastrous impact on most of the practices represented here today."

It appears that the pendulum has begun to swing the other way and we can expect to see more accreditations in the future. In the first place, proposals now require a certain number of engagements per year in the specialty, rather than a specific number of hours of practice. That means smaller firms will have an easier time meeting the requirement. But the much more important change that has made further accreditations so much more palatable to the average CPA is the seminal change that has occurred in the profession itself. As competition increased, and as CPAs diversified into new areas, even those who had initially been opposed to new accreditations began to recognize that an accreditation in a particular area of expertise would give CPAs a legitimacy in the eyes of the public that others could not match.

The changing views of Robert Israeloff exemplified this attitudinal shift. Bob was AICPA chairman from 1994 to 1995, and is a principal in the midsize firm, Israeloff Tratner & Company. He was initially in the forefront of opposing new certifications, but in 1995 he agreed to become chairman of the special committee on specialization. As the business environment evolved, so too did the views of Bob and others.

"What once was a public interest argument in favor of certifying different specialties has now become a self-interest argument," he explains. "If the profession is moving toward consulting services rather than traditional accounting, it is in the interest of the profession to stake out a claim of expertise."

In recommending that additional specialties be accredited, the special committee began to focus on specialties for which there might be a public need for assurances of competency. As the population ages, and as many baby boomers become at least partially responsible for the care of their parents, there's an expanding need for such services as estate planning and issuing assurance reports on particular nursing homes.

SPECIAL COMMITTEE ON ASSURANCE SERVICES

Another effort during these years to explore an expansion of CPA services was made by the Institute's Special Committee on Assurance Services, formed as my AICPA tenure neared its end. The committee was an outgrowth of an AICPA-sponsored conference held in Santa Fe in 1993, at which representatives of small and medium-sized firms, regulators, and scholars concluded there was a need to expand the role of the auditor. There was a growing realization that revenues from traditional accounting and audit services had been largely flat during the previous decade, while consulting services had been growing at about 10 percent annually. Nonaccountants, unconstrained by the accounting profession's rules, regulations, and standards, were offering similar consulting services as accountants, sometimes at a competitive advantage. Instead of simply reporting on the reliability of assertions by management (usually financial statements), perhaps there was also a need for auditors to comment on the reliability or relevance of other assertions by management. This would considerably expand the audit function.

The special committee, dubbed the Elliott Committee after its chairman, Robert K. Elliott, a senior partner at KPMG Peat Marwick who became AICPA board chair in 1999, was established "to develop new opportunities for the accounting profession to provide value-added assurances." As Bob noted in a symposium held on January 5, 1996, sponsored by the CPA Journal,

> Information technology will expand the types of information and the ways it can become available. The question is who is going to provide the assurances— the profession, or the new technology-savvy competitors? Are we going to reengineer the profession so that we can take advantage of these opportunities? That's what this effort is all about.

> The Special Committee is involved in the transformation of the public accounting profession now under siege by intense competition and its failure to keep pace with change as society moves from an Industrial Age to the Information Age. Some industries and companies have succeeded in making the changes necessary; others have not. Can a profession such as accounting, with a financial accounting system going back to Pacioli in the 15th century and an auditing system that dates back to the end of the 19th century, ever hope to keep pace with other industries with product life cycles now being measured, in some cases, in months? That is the question we hope to answer.

The Elliott report concluded that given the limited opportunity for future growth of traditional accounting services, most notably the audit, new activities that create value for business decision makers had to be developed to ensure that the profession remained relevant. At the May 1995 Council meeting, Bob noted that the audit function had generated flat revenue during the previous six years, while gross domestic product had been increasing. What's more, the number of people doing audits had declined by about 20 percent, partly because litigation-related costs had made auditing a relatively unattractive business in which to be involved.

Audits of financial statements had also lost some of their relevancy, as investors began to use sources of information not found in the typical annual report. The reason stock prices are generally unaffected by published financial statements is that quarterly financial statements, press releases, and analyst meetings have already disclosed the information, so they have already been taken into account by investors.

Another criticism of the old-style financial statement was that one size fits all no longer met the needs of investors, management, creditors, the SEC, and the IRS. Rather, it is a compromise that completely satisfied no one.

The Elliott committee suggested a number of opportunities that the profession could take to expand its services by leveraging our existing skills. The good news, the committee reported, was that the positive public perception of the profession provided a strong base on which to build these new services. The profession's competitive advantages included its reputation for independence, access to clients, concern for the public interest, and the quality control systems firms already had in place. The committee suggested that while accountants have historically provided financial statements based on historical, cost-based information, perhaps customers might prefer:

- Real-time, continuous data
- Prospective data
- Value-based data
- Relevant nonfinancial data
- Statements capable of extracting only the information that the client wanted

One of the central themes of the committee's final report was how important it was to keep up with technology, which was accelerating a shift in power from the producer to the consumer. It predicted that data access, largely through the Internet, would replace periodic reporting, and that electronic commerce would grow substantially. The report foresaw a major role for information technologies in developing new assurance methods, with accountants using electronic networks to collect and distribute information.

GLOBALIZATION

One issue we did not identify in the initial 1984 future issues report, but which was subsequently added to the first strategic planning report, was globalization. By the mid-1980s, some multinational corporations had more economic influence than many countries. These global companies had become powerful entities unto themselves, for which national boundaries had little meaning. This was a situation unparalleled in history. Globalization received another powerful boost in 1989 with the fall of the Berlin Wall and the transformation of eastern European economies from communism to capitalism. Suddenly there were large industrialized countries in serious need of Western accounting expertise.

Internationally, this situation had profound political ramifications. With sovereign economies intertwined like never before, and as armed conflict became anathema to the major international corporations, a greater stability settled into the world economy. The reality of a single economy in which a stock market crash in Asia instantly affects the New York Stock Exchange's Dow Jones Average was fueled by the instantaneous communication of information. One man stands in front of a tank in Tiannamen Square and the world is watching. Princess Diana is killed in an automobile accident in an underground tunnel in Paris and hundreds of millions of people worldwide are on the scene via television and the Internet.

How did all this relate to the accounting profession? As more and more companies did business globally, even small firms serving small clients had to be concerned with what occurs on a worldwide basis. Currency issues, the derivatives phenomenon, difficulties in Asian economies, and stock markets worldwide all affect a surprisingly large number of companies. At the same time, global companies began operating in many different countries simultaneously, which means they have to comply with the accounting and financial reporting requirements of all the nations in which they do business. (See Chapter Six.)

BIG EIGHT, BIG SIX, BIG FIVE, WHAT'S NEXT?

Consolidation was another issue that the Future Issues Committee predicted would help transform the profession. Indeed, as recently as 1985, the term "Big Eight" was firmly entrenched as the way to describe the largest accounting firms: Arthur Andersen; Arthur Young; Coopers & Lybrand; Deloitte, Haskins & Sells; Ernst & Whinney; Peat, Marwick, Mitchell; Price Waterhouse; and Touche Ross. But then Ernst and Whinney merged with Arthur Young, and Deloitte, Haskins & Sells combined with Touche Ross. During the summer of 1998, Price Warehouse and Coopers & Lybrand joined, thereby displacing Andersen Worldwide as the world's largest accounting firm.

In part, this massive consolidation represented an effort at one-upmanship. Everyone wants to be the biggest and the best. But more importantly, it was also an attempt by the largest firms to provide a full range of services to their global clients, with improved geographic and strategic coverage of their needs. As companies became international entities, they needed international accounting and consulting firms to support them in the global environment.

As a result, large companies suddenly had only a small handful of firms from which to choose. Small CPA firms simply do not have the human resources, expertise, or global reach to audit Fortune 100 companies. Competition among these five firms is still fierce, so from a traditional antitrust perspective the most recent mergers would not seem to be a problem. U.S. and European regulators, however, would undoubtedly look closely at any further consolidation.

LOOKING FORWARD

I like to believe that our emphasis on forward planning during these years replaced a more event-oriented approach of managing the evolving issues affecting the accounting profession, and that the profession was better off as a result. But as subsequent chapters will demonstrate, some events, such as congressional scrutiny and the liability crisis, produced such a firestorm that no amount of planning could completely prepare us for the onslaught.

Chapter 2

Sue the Accountants: Fraud Detection and Legal Liability

Imagine for a moment you're a CPA auditing the books of a multimillion dollar insurance company. Your audit raises no doubts about the information that's been provided to you, and the company appears to have followed generally accepted accounting principles in preparing its financial statements. True, income from premiums is sharply higher than in past years, but on closer inspection it becomes evident that the company did a good job of laying off their risk on other insurers. Besides, you have found no reason to doubt the integrity of the top management of this company, or the reliability of their financial information.

Or perhaps you're auditing a well-known New York City retail electronics firm famous for its flamboyant television advertising. It's the rare New Yorker who doesn't identify this retailer with low prices, so you are not surprised that profits for the year have jumped to more than $20 million.

Or maybe you're inspecting the books of a disk drive manufacturer, another company with a solid reputation in a fast-growing industry that as far as you can reasonably determine has followed generally accepted accounting principles.

But what if it turns out that the insurance company's employees have been colluding to make its financial statements look good by "selling" insurance to fictitious people, reinsuring them with other insurance companies to collect fees, and then even "killing them off" to make the operation appear legitimate? Or what if the chairman of the electronics store has created an intricate web of phony documents in order to overstate income? And what if the disk drive company is shipping bricks instead of disk drives, counting

21

them as sales, and then taking them back as returns in order to overstate inventory and understate the cost of sales?

What happens is that eventually there is a day of reckoning. No matter how crafty the employees perpetuating the reporting of fraudulent information, no house of cards is sturdy enough to remain standing forever. No company can sustain fictitious profits indefinitely. The fraud is inevitably revealed, shareholders lose millions, and often the company's only recourse is bankruptcy, which in fact occurred within a 30-year span at all three companies described above—Equity Funding Corporation of America, Crazy Eddie, and Miniscribe.

WHERE WERE THE AUDITORS?

And then come the lawyers. A bankrupt company will typically be unable to pay its creditors, much less financially reimburse the thousands of shareholders who have lost money as the company's stock price evaporates. Since the people who actually perpetuated the fraud are unlikely to have the financial wherewithal to satisfy any legal claims, litigants begin searching elsewhere for someone to sue. "Where were the auditors?" they ask, and often they decide that the accounting firm, or at least its insurers, represents the only deep pocket left standing.

This is just what occurred at an unprecedented pace during my tenure as AICPA president. Compounded by the recession of the 1970s which provoked a record number of bankruptcies, the savings and loan crisis, and an overall explosion in the number of litigants and class action suits, accounting firms paid out enormous sums as a result of lawsuits by shareholders and other injured parties.

In an environment looking for a scapegoat, within a society that had become the most litigious in the history of civilization, it was no surprise that some legal judgments were difficult for the profession to swallow. After Miniscribe's bankruptcy, for example, in February 1992 a civil jury found Coopers & Lybrand liable for $200 million. But investors had lost only $13 million. In December of that same year, Ernst & Young paid the government a record $400 million to settle charges that its auditors contributed to the S&L debacle.

JOINT AND SEVERAL LIABILITY

The accounting profession's most onerous legal exposure was due to the prevailing "joint and several" liability rule, under which a plaintiff could collect all or part of his or her damages from any defendant found liable, regardless of that defendant's proportionate fault. In other words, even if an auditor were found to be only 10 percent responsible, the auditor could still be required to pay 100 percent of the awarded damages. The rule encouraged plaintiffs to name as many "deep pocket" defendants as possible, regardless of the degree of that particular defendant's actual responsibility.

To the accounting profession, this seemed patently unfair. It meant that it was often more prudent to settle even the most frivolous lawsuit than to risk a judgment that could run into the tens or even hundreds of millions of dollars. Often a settlement simply made financial sense, if only to avoid the certainty of expensive legal costs. Indeed, according to a report by the then Big Six entitled *The Liability Crisis in the United States: Impact on the Accounting Profession*, in 1991 the average settlement by a Big Six firm was $2.7 million, compared with the average legal cost per claim of $3.5 million. Many firms also were reluctant to suffer the damage to their reputation that inevitably resulted from a public legal judgment.

But whether through settlements or judgments, the cost to the profession was enormous. According to a 1996 report by the Arthur Andersen Public Review Board, in 1991 the direct costs of litigation for the Big Six firms in the United States was $367 million, or approximately 7 percent of accounting and auditing revenues. Three years later it had tripled to more than $1 billion, or to about 20 percent of revenues. This was not just a Big Six problem. At least two second tier firms, Laventhol & Horwath and Pannell Kerr & Forrester, were forced into bankruptcy in the early 1990s, partly because of their liability exposure.

LIABILITY INSURANCE SKYROCKETS

Given these numbers, it's no surprise that insurance costs for the accounting profession made a quantum jump as well. The rule of thumb used to be that firms had to pay $1000 for $2 million of coverage. That ratio remained steady from 1972 through 1983, without even an inflationary increase. But on November 1, 1984, the AICPA's insurance program's administrators were notified by their lead underwriters, Crum & Foster, that our reinsurers were demanding a 100 percent increase in premiums, a doubling of deductibles, and certain other changes such as the engagement of an outside independent consulting firm to study the viability of the Institute's entire insurance program.

According to *The Liability Crisis* report, insurance for 96 percent of accounting firms with more than fifty CPAs rose 300 percent between 1985 and 1992, with deductibles rising nearly sixfold. And between 1980 and 1986, under the AICPA's own liability insurance plan the premium for $1 million in coverage for firms with at least twenty-five professionals skyrocketed from $64 per person to $1160 per person, while deductibles doubled.

To some extent at least, the ultimate effect of higher insurance costs was to redistribute these added costs, since the only way accounting firms were in a position to pay higher premiums was to reduce partners' compensation or increase the fees they charged clients. In the latter case, the accounting firm almost became an intermediary as it passed on its liability costs to clients.

Insurance premiums became so high so quickly that some firms decided to operate without any insurance at all, knowing full well that a sizable judgment had the potential of bankrupting all but the very largest firms. Other firms obtained insurance by contributing to captive mutual insurance companies formed to provide professional liability insurance. The California state society, for example, developed a very successful captive insurance company for its members which continues to this day.

GROPING FOR SOLUTIONS

Certainly there were lawsuits in which CPAs were culpable. In rare cases, CPAs were part of the fraud, and certainly in other instances, particularly with the benefit of 20-20 hindsight, accountants were less than diligent in detecting impending financial problems. In many of these cases, injured parties were justifiably entitled to compensation from the accounting firm. Indeed, in February 1989 the General Accounting Office (GAO) reviewed the audits of eleven S&L associations in the Dallas Federal Home Loan Bank District and concluded that for six of them, CPAs did not properly audit or report financial or internal control problems in accordance with professional standards.

But the important question we had to ask ourselves during these litigious years was this: Even when auditors had done their job well and when the company in question had complied with generally accepted accounting principles, was that enough in our complex, litigious society to protect the accountant from liability if the client later suffered a financial decline? In the period between 1980 and 1995, often it was not. But just what *was* an accountant's responsibility for detecting fraud, particularly when faced with a conspiracy to deceive on the part of top management? What were an accountant's professional responsibilities, and what steps was the profession taking to improve the audit, on the one hand, and reduce frivolous suits, on the other? The profession set out during these years to slow the liability onslaught by embarking on two parallel paths: by improving the audit process and by obtaining liability relief.

In the first place, we looked within and realized there were actions we could take to improve the auditor's ability to detect fraud or other financial problems. We determined, for example, that the financial information public companies were obliged to disclose needed revamping, and that auditors needed to be reminded that one of their responsibilities was to find material misstatements, whether caused by fraud or error. We needed to clarify what was expected of an audit, and to provide the auditors with the skills necessary to meet those expectations.

Sometimes our efforts were based on simple common sense. At our May 1986 Council meeting, for example, Marvin Stone, former chairman of the board, pointed out that his own practice was devoted almost entirely to litigation consulting. "In recent years, I'm sorry to say, that usually means defending CPA firms being sued for alleged malpractice," he told us. "I cannot urge all of you strongly enough to be better at screening the clients you take on, particularly those which have made a public offering. Almost always the kind of case I get involved in are ones in which the firm's client was a bad apple, and investors have turned to the deepest pocket they can find to pay the unfortunate investors. Many times, you can avoid all this by more carefully investigating prospective clients."

But in addition to the steps we could take ourselves, or at least heavily influence, the profession also believed that an equally important task was to lobby both the courts and Congress to obtain some sort of relief from a legal system run amuck. By the mid-1980s, federal and state laws concerning an auditor's exposure to liability had become a hodgepodge of conflicting precedents and multimillion dollar exceptions. Joint and several liability was one of our grievances, but there were others. Class action suits had proliferated to such an extent and with such intensity that it was the rare securities liability lawsuit that didn't add an accounting firm or two to its long list of defendants. What's more, firms were increasingly being sued under the Racketeer Influenced and Corrupt Organizations Act (RICO), something we believed Congress never had intended. Passed in 1970 to combat organized crime, RICO was increasingly being used to sue legitimate businesses, including individual CPAs and accounting firms. It also allowed plaintiffs to seek treble damages and reimbursement of all legal costs. (See Chapter Three.)

LAVENTHOL & HORWATH BANKRUPTCY

By 1990, multimillion dollar lawsuits, skyrocketing insurance costs, and a legal environment that had everyone confused as to the responsibilities of the accountant for detecting financial problems, much less fraud, had the entire profession on edge. And then just before Thanksgiving 1990 an event occurred that all of us had been dreading but dared not speak aloud—the bankruptcy of a major accounting firm due largely to its liability exposure.

During the high-flying 1980s, the seventy-five-year-old Philadelphia firm of Laventhol & Horwath nearly quadrupled its revenues to more than $345 million and expanded to more than fifty offices, in the process becoming one of the ten largest accounting firms in the United States. Much of its growth had come from acquiring smaller firms and expanding into nonaudit services, mostly in various consulting areas. At Laventhol & Horwath, that meant focusing on such specialties as the restaurant and lodging business, to the point of establishing an interior decorating department.

Although far afield from traditional accounting, these new services proved extremely lucrative. In the process, however, Laventhol & Horwath also probably took on clients it shouldn't have, including such headliners as televangelist Jim Baker. After Baker was sentenced to prison, former members of the PTL television ministries sought $184 million from Laventhol & Horwath, whom they accused of helping Baker appear solvent when his operation was in fact in technical default.

But the PTL case was only Laventhol & Horwath's most publicized problem. In 1988 it became the first firm to lose a jury trial under RICO, and in 1989 it paid out $57 million to settle various lawsuits accusing it of shoddy auditing. It also had a startling number—eighty, by one estimate—of cases pending before the courts.

Laventhol & Horwath's bankruptcy sent a chill throughout the accounting profession. Suddenly no one was safe. If liability problems had caused the demise of a top ten accounting firm, what was to prevent the same thing from happening to the sixth largest, or for that matter the first?

The threat to the profession was made all the more real by the 3500 Laventhol & Horwath employees, including 350 partners, who were suddenly on the street looking for work. Even those partners who immediately took over their local offices and managed to retain many of their clients were not feeling safe, since all partners were being held personally liable for the financial claims of the firm's creditors. While a corporate bankruptcy filing protects a company from its creditors, Laventhol & Horwath was a partnership, which meant the partners could be sued individually. Indeed, some Laventhol & Horwath partners—even some *retired* partners—were apparently forced into personal bankruptcy despite the fact they were in no way involved with the audits that were the subject of litigation.

The profession recognized that the Laventhol & Horwath's bankruptcy was by no means an isolated case of one firm extending itself too far and paying the consequences. Rather, we realized that it was, unfortunately, symptomatic of the problems facing us all.

FORM OF PRACTICE

One immediate result of the Laventhol & Horwath bankruptcy was that accounting firms began to reorganize into limited liability partnerships (LLPs) so that only the partners specifically involved with the wrongdoing would have to use their personal assets to pay any judgments. The firm would be fully liable, but under an LLP the personal assets of partners not involved in the case would be protected.

In the early days of the profession, firms routinely operated as limited liability corporations, but in 1938 the AICPA's membership voted to prevent anyone affiliated with such a corporation from being a member. The concern was that these accountant-corporations were being formed for personal reasons rather than for the best interests of the public, and that members were concealing their identity behind these corporate shields.

But by the 1980s, Rule 505, our strict prohibition of professional corporations, had become an anachronism. In fact, we could find no other national professional association that restricted the form in which its members practiced their professions.

In December 1989, the Institute's board of directors appointed a task force to investigate the form of practice issue and offer its recommendations. The task force found wide support for lifting the prohibition, and at the May 1991 Council meeting the following resolution was passed:

RESOLVED: Under Rule 505, a member may practice public accounting in a business form other than a proprietorship or partnership only if such entity has the following characteristics:

Ownership. All owners shall be persons who hold themselves out to be certified public accountants or public accountants and who perform for clients types of service performed by certified public accountants or public accountants. Owners shall at all times own their equity in their own right and shall be the beneficial owners of the equity capital ascribed to them.

Transfer of Ownership. Provision shall be made requiring any owner who ceases to be eligible to be an equity owner to dispose of all his or her interest within a reasonable period to a person qualified to be an owner or to the entity.

Directors and Officers. The principal executive officer shall be an owner and a member of the entity's governing body. Lay governing body members shall not exercise any authority whatsoever over professional matters.

Conduct. The form of organization shall not change the obligation of its owners, directors, officers, and other employees to comply with the code of Professional Conduct established by the AICPA.

The Council then passed a resolution authorizing a mail ballot of the membership "seeking to modify present Rule 505—Form of Practice and Name to make it possible for CPAs to practice in any form of organization permitted by state law or regulation." It was subsequently ratified by membership. The AICPA also lobbied state legislatures to allow accounting firms to reorganize in this manner, so that today LLPs are allowed in every state except Wyoming and Vermont. Many states also allow other forms of incorporation, such as limited liability and professional corporations.

TREADWAY COMMISSION

Restructuring the legal incorporation of accounting firms was a relatively superficial, stopgap measure aimed at curbing our liability exposure. We understood perfectly well that more fundamental solutions were needed. At the core of this issue was that despite our protestations, the public, and therefore juries, couldn't shake the idea that an unqualified audit report was the accounting firm's equivalent of a "warranty" of sorts. Their perception was that it somehow guaranteed that all the financial information was accurate, or even that the company was a good investment. We set out to change these perceptions, and at the same time improve the audit process in order to increase the likelihood that an audit would detect fraud.

The most important step the Institute undertook toward this end was the appointment in October 1985 of the National Commission on Fraudulent Financial Reporting. Chaired by James C. Treadway, former SEC commissioner and later Paine Webber general counsel, the Treadway commission was charged with identifying factors that led to fraudulent financial reporting and to recommend steps that would reduce its incidence.

Appointment of the commission was the brainchild of Ray Groves, chief executive of Ernst & Young, and chairman of the AICPA board of directors between 1984 and 1985. Ray's motivation came from a belief that much better coordination was needed between the internal and external audits.

"We appointed Treadway out of a conviction that the totality of financial statements involved more players than just the outside auditor," Ray recalls. "We didn't think people understood that. Our purpose was not to take the responsibility away from ourselves, but to make the other people involved—those who worked inside the companies being audited—realize that they too had a vested interest in producing accurate financial information."

It was also Ray Groves's inspiration to find a well-respected, independent person to chair the commission. As SEC commissioner, James Treadway had expressed the most concern of any of his colleagues about financial statements, so when he announced he was leaving the SEC, Ray asked him to chair the commission we were in the process of establishing. The direction we wanted to take, as spelled out by Ray, intrigued Treadway, and fortunately he agreed to our request.

The Treadway commission's objective was to investigate the possible causes of fraudulent financial reporting, defined as "intentional or reckless conduct resulting in materially misleading financial statements." In an effort to come up with meaningful recommendations, commission members, its staff, and outside consultants spent thousands of hours investigating how specific instances of fraud might have been influenced by corporate pressures, ineffective internal controls, failures in auditing standards or professional conduct, or a lack of independence or education of the auditors.

After two years of work, the Treadway commission issued forty-nine specific recommendations to deter fraudulent financial reporting. Most of them, I'm happy to say, were subsequently implemented either through congressional legislation, regulatory action, or by the profession itself.

Some of the recommendations were directed toward the top management of public companies, such as creating an environment conducive to fraud prevention. That included the strict enforcement of a written code of corporate conduct and having an independent audit committee of the board of directors vigilantly oversee the company's internal reporting process.

Many other recommendations were directed at the independent public accountant, including specific changes in generally accepted auditing standards (GAAS) to better recognize the auditor's responsibility for detecting fraud, and the establishment of a peer review process. The Treadway commission also recommended that a CPA's education—at both the examination and CPE levels—emphasize the knowledge, skills, and ethical values necessary to understand and expose fraudulent financial reporting.

The commission's most fundamental conclusion was that both company personnel *and* the outside auditors shared a responsibility for preventing and detecting fraudulent financial reporting. On the one hand, it recommended that GAAS be changed to better recognize the independent public accountant's responsibility for detecting fraud. But on the other hand, it also emphasized the often overlooked importance of public companies having their own internal controls aimed at preventing and detecting fraud.

COMMITTEE OF SPONSORING ORGANIZATIONS OF THE TREADWAY COMMISSION

After the Treadway commission issued its report, its five sponsoring organizations established a committee to monitor compliance with its recommendations. Former board chairman Mike Cook agreed to chair the committee, referred to as COSO, the acronym for Committee of Sponsoring Organizations of the Treadway commission. In addition to the AICPA, the sponsoring organizations included the American Accounting Association (AAA), the Financial Executives Institute (FEI), the Institute of Internal Auditors (IIA), and the National Association of Accountants (NAA).

In 1992, COSO issued a report entitled *Internal Control—Integrated Framework*. It sought to bridge the gap between management's emphasis on control and the auditor's concern about preventing or detecting fraud or errors. It reaffirmed Treadway's conviction that a public company's top management must itself take responsibility for overseeing the financial reporting process by identifying, understanding, and assessing factors that may cause the company's financial statements to be misstated.

The COSO report also endorsed the Treadway commission's recommendations that public companies develop and enforce a written code of corporate conduct to be reviewed annually by a company's audit committee. It cited a survey it had conducted of 8500 public companies revealing that 87 percent of companies with more than 10,000 employees, but only 60 percent of companies with between 1000 and 10,000 employees, had a written code of conduct. Only 54 percent of companies had a separate internal audit function, although 98 percent of companies with more than 10,000 employees and 80 percent with between 1000 and 10,000 employees had such a written code.

At the October 1988 Council meeting, Cook laid out COSO's preliminary recommendations, noting that "the credibility of public financial reporting is critical to us if we are going to maintain public confidence in the work of our profession and in the financial reporting process. Fraudulent financial reporting is relatively rare, but it is nonetheless a very serious problem in terms of losses to innocent victims and of the serious damage it does to our credibility. While there are only a handful of these cases out of the 11,000 or so audits of public companies done every year, the 99.4 percent purity standard just isn't good enough. We have to work very hard to seek perfection, recognizing that perfection is never possible in a human endeavor."

The final COSO report had more than fifty recommendations aimed at corporate management and their audit committees, internal and external auditors, law enforcement and regulatory authorities, and the educational community. Mike reported to the Council that following through on these recommendations would be a key ingredient in our effort to demonstrate effective self- regulation and convince Congress that there was no need for further government control over the financial reporting process. "If this report becomes a shelf document to gather dust as some of its predecessors have, the consequences for us I believe would be very serious," he warned. "We know Congress is watching."

KIRK REPORT

An important study during these years was conducted in 1993 by a three-member advisory panel on auditor independence appointed by the Public Oversight Board (POB; see Chapter Four). It was chaired by Donald Kirk, a founding member and former chairman of the FASB and a former partner at Price Waterhouse. The other two members were George Anderson, chairman of the Special Committee on Standards of Professional Conduct (see Chapter Four), and Ralph S. Saul, a past president of the American Stock Exchange.

The panel was charged with determining whether steps needed to be taken to improve the objectivity, integrity, or independence of auditors. It interviewed close to 100 professionals in an effort to better understand the problems of maintaining independence in a new auditing environment made up of a multitude of new services accountants were performing.

Two years later, *Strengthening the Professionalism of the Independent Auditor* (informally known as the Kirk report) concluded that there was no need for further rules or regulations to strengthen the conflict-of-interest aspects of auditor independence. It declared that in large public auditing firms, the auditing function was inherently different from other services accounting firms provided. The Kirk report did suggest, however, that stronger and more accountable boards of directors would help strengthen auditor independence. The report affirmed that it was the responsibility of boards of directors and their audit committees to look after the interests of stockholders. It recommended that board members listen more closely to outside auditors concerning the appropriateness of the accounting principles being used.

The panel also suggested that auditor independence could be further encouraged by discouraging a close relationship between corporate management and auditors in determining both accounting principles and audit fees. Ultimately, the panel noted, boards of directors must focus more on the quality of a company's financial reporting than on its acceptability by auditors.

The Kirk report concluded with a plea for a more congenial and less adversarial relationship between the SEC and the accounting profession, particularly with respect to providing relevant information to the public. In the panel's opinion, the SEC should see itself only as the arbiter of last resort, and then only after much investigation and due care had been taken to mediate any disputes. While the SEC certainly had legitimate watchdog responsibilities, it was also its responsibility to reduce the threat of unwarranted litigation targeting the accounting profession.

OTHER STUDIES TARGETING LIABILITY CONCERNS

In addition to Treadway, COSO, and Kirk, a host of other studies and reports were issued during these years that joined the effort to reduce the expectation gap and improve the liability situation. In 1985, for example, just as the Treadway commission was getting started, Price Waterhouse (PW) weighed in with three proposals of its own:

1.   The expansion of auditing standards so that the audit would be better equipped to search for management fraud

2. The creation of a statutory self-regulatory organization (SRO) that would give the public increased assurances about the quality control and peer review process through which the profession assesses levels of performance, independence, and adherence to auditing standards
3. The development of a program to seek liability relief

Many in the profession, including the other Big Eight firms, were particularly concerned about that part of the PW proposal calling for the creation of an SRO to regulate CPA firms practicing before the SEC. We felt that the existing body of regulation was sufficient, and that a new SRO would only duplicate and undermine the self-regulatory activities already being performed within the private sector. We were also concerned that once a legislative process got started, there would be no assurance that it would be enacted along the lines recommended by PW.

The following year, the other large accounting firms submitted their own eight recommendations to the AICPA board of directors aimed at "improving the relevancy, reliability and credibility of financial information through greater disclosure concerning risks and uncertainties in financial statements, independence, and peer review." That report simply ignored Price Waterhouse's SRO proposal, which never received significant support.

RISKS AND UNCERTAINTIES

At the same time Treadway was studying ways the audit could maximize the detection of fraud, the Institute also established a separate task force on risks and uncertainties. Its goal was to suggest how financial reporting could, when appropriate, alert users to the possibility that a public company was in danger of failing, or even experiencing a severe financial setback. A number of this task force's recommendations were subsequently adopted, including:

- Additional disclosures in financial reports such as an explanation about the kinds of products or services the enterprise sells and in which markets they are sold
- A discussion of any significant estimates used by management to measure assets and liabilities
- Information about vulnerability due to concentrations in the enterprise's assets, customers, or suppliers

All these changes were generally well received by both preparers of financial information and users. Ironically, however, the new rules did little to limit an accounting firm's potential liability. In fact, the profession's avowed effort to uncover fraud in some instances may have actually served to exacerbate the liability crisis, since for the first time auditors seemed to be offering their assurances that the financial information found within an annual report was free of fraud. If it turned out that top management of a company had pulled the wool over everyone's eyes, we were potentially putting ourselves in an even more vulnerable position when faced with a lawsuit.

AUDIT COMMITTEES

One conclusion that almost all these many committees and reports had in common was the importance of corporate audit committees. Regardless of the vigilance on the part of the individual auditor, the profession, and the federal government, the most effective way to prevent fraud and sound the alarm before problems occur was through an effective, watchful audit committee. This is not a revolutionary concept. At various times, the Treadway commission, the POB, and the SEC pointed to a strong audit committee as perhaps the single most effective tool for improving the effectiveness of audits. Professional groups in other countries also recognized their importance, including the Macdonald Commission in Canada and the Cadbury Committee in the United Kingdom.

The Treadway commission had also touched on these issues. It had found that a significant proportion of companies cited in actions by the SEC had audit committees, so the commission had already made the point that simply having an audit committee was not enough of a safety net. In order to operate efficiently, audit committees must be independent, informed, and vigilant.

A survey conducted by Dorothy A. McMullen, an assistant professor of accounting at Rider University in New Jersey, and K. Raghunandan, associate professor of accounting at the University of Massachusetts, examined 51 companies that had had some kind of financial reporting problem between 1985 and 1989. Either the SEC had instituted an enforcement action against them, or they had been forced to issue a restatement of quarterly results. For comparison purposes, a random sample of 100 companies with no financial reporting problems was also studied. The study reinforced the Treadway commission's recommendations that to be effective, audit committees should have only outside directors. It found that of the companies with financial reporting problems, 67 percent had audit committees composed of outside directors, compared with 86 percent of those companies without financial reporting problems.

Problem companies were also much less likely to have CPAs on their audit committees. Of the companies with financial reporting problems, only 6 percent had an audit committee with at least one CPA, compared with 25 percent of those companies without problems. The study also emphasized the desirability of audit committees conducting regular meetings. It found that only 23 percent of audit committees of problem companies had regularly scheduled meetings three or more times per year, compared with 40 percent of audit committees within companies without financial reporting problems.

Other reports confirmed many of these results. The GAO's 1991 study, *Audit Committees: Legislation Needed to Strengthen Bank Oversight*, found that many failed financial institutions had audit committees whose members lacked banking or financial management expertise. As a result, the 1991 Federal Deposit Insurance Corporation Improvement Act required audit committee members of large banking institutions to be knowledgeable about financial and banking-related matters.

Price Waterhouse, in its 1993 internal report called *Improving Audit Committee Performance: What Works Best*, declared that the key to audit committee effectiveness was the members' expertise in accounting, internal controls, and auditing. And Coopers & Lybrand, in its 1994 study of audit committees, which they turned into a widely distributed *Audit Committee Guide*, recommended that audit committees meet three or four times a year.

INTERNAL AUDITS

As discussed, the Treadway commission directed a number of its recommendations toward public companies, urging that internal controls be strengthened and that there be much more consistency between the internal audit and the external independent audit. Internal accountants concern themselves not only with external reporting requirements, but, more importantly, with a wide variety of financial information. This allows company management to operate their business more effectively by establishing systems of internal control for measuring and validating information. The most innovative of these processes can give companies a distinct competitive advantage.

There is an entire discipline—management accounting—concerned with the accumulation and interpretation of information used to make all kinds of internal decisions—everything from pricing, to evaluating research and development expenditures, to deciding whether to expand the work force. Internal auditors test the effectiveness of the systems. Their ultimate goal is to establish a reporting system that will satisfy management, users, and external auditors, and also create a process that will most efficiently run their company.

Internal accountants and auditors even have their own organizations. If they are CPAs, they may well belong to the AICPA, but they may also belong to the Institute of Management Accountants, the Financial Executives Institute, or the Institute of Internal Auditors.

ACCOUNTING AND AUDITING STANDARDS

While the AICPA leadership applauded any and all efforts to strengthen the internal audit function, we also knew it was our responsibility to do all we could to strengthen the external audit. Remember that in addition to requiring independent audits, the securities acts of 1933 and 1934 also gave the Securities and Exchange Commission (SEC) the authority to set financial reporting standards for publicly held companies. In the 1930s, there was serious discussion in Washington about whether the federal government should itself establish standards for preparing financial statements of publicly held companies. But the Institute, aided by the state societies, persuaded the SEC that the authority should be left in the hands of the private sector. Since 1973, standards have primarily been set by the Financial Accounting Standards Board (FASB), an independent body that sets the ground rules for accumulating and reporting information.

Accounting standards cover a broad range of topics—everything from broad concepts such as revenue recognition, to more specific rules such as how to report information about the company's different businesses and to ground rules for measuring and/or disclosing "risks and uncertainties." A section called "contingencies" is also included in the footnotes of the financial statement to describe a particular litigation or other special circumstance that could possibly have an adverse effect on the financial performance of the organization.

Because business transactions became so complex, and because financial statements were now expected to summarize such large amounts of information, it often became necessary to supplement the data with footnotes that provided more detail. These footnotes explain more fully the particular accounting alternatives selected by a company. In the case of publicly held companies, management is also required to spell out information about revenue, expenses, and liquidity in a separate section of the annual report called "management's discussion and analysis."

While the securities statutes vested the SEC with the authority to set accounting standards, it is more problematic whether the SEC has the authority to set "generally accepted auditing standards." The position of the AICPA over the years has been that the auditing profession, rather than the SEC, has the ultimate authority to establish these standards. The alternative would be to allow the federal government to use the audit process to control the way business is conducted, thereby contradicting some very basic tenets of the free enterprise system. So instead, the Auditing Standards Board (ASB), an entity within the AICPA, sets the ground rules for how an auditor determines whether the information reported is reasonable and whether it conforms with generally accepted accounting principles. As a practical matter, however, before promulgating auditing standards, the ASB carefully considers the views of the SEC's Chief Accountant, as well as the views of many others invited to comment on the proposals.

AUDITING STANDARDS BOARD

The first auditing standards were actually issued in 1939 when the AICPA's predecessor organization, the American Institute of Accountants, authorized the appointment of a standing committee on auditing procedure. Two years later this committee issued a *Statements on Auditing Procedure,* designed to guide the independent auditor in determining auditing procedures. In 1951 the committee issued the *Codification of Statements on Auditing Procedure* that consolidated the first twenty-four of these pronouncements.

In 1947, after the SEC adopted the requirement that independent auditor's financial statements comply with GAAS, the committee submitted a report entitled *Statement of Auditing Standards—Their Generally Accepted Significance and Scope.* It wasn't until twenty-five years later, however, in November 1972, that all the committee's pronouncements were codified into a single presentation. The name of the committee was simultaneously changed to the Auditing Standard Executive Committee (AudSEC) as a recognition of its role as the AICPA's senior technical committee charged with interpreting GAAS.

Between 1976 and 1979, while still a partner with KMG Main Hurdman (since merged into KPMG), I served as chairman of AudSEC. So I was still chairman when in 1978 the ASB was formed as the successor to the executive committee. In the process, the new board was charged with promulgating auditing standards and procedures that:

1. Defined the nature and extent of the auditor's responsibilities
2. Provided guidance to auditors in carrying out their duties and enabling them to express an opinion on the reliability of the financial information
3. Made special provision, where appropriate, to meet the needs of small enterprises
4. Took into account the costs standards impose on society in relation to the benefits reasonably expected to be derived from the auditor function

So while accounting standards for financial reporting are for the most part promulgated by the independent FASB, auditing standards are still set directly by the profession—by the ASB, an arm of the AICPA. The ASB gets its authority from the long-held acceptance by regulators, the accounting profession, and the public that it is the definitive source of auditing standards, as well as from the fact that the SEC has publicly acknowledged the ASB's role in these matters. This has effectively given the ASB's pronouncements the force of law, and they are recognized by the courts as the standards applicable to audits of financial statements.

AUDITOR'S STATEMENT

The independent auditor's application of GAAS is indicated in the annual report of every public corporation (and many nonprofit organizations) on the page entitled *Report of Independent Public Accountants*. It states that the goal of the audit is "to obtain reasonable assurance that the financial statements are free of material misstatement" and that the audit was performed "in accordance with generally accepted auditing standards."

Let's take a closer look at these words because they tell a lot about the audit function. In the first place, the word "independent" in the title, *Report of Independent Public Accountants*, is very telling. It wasn't until the securities acts of 1933 and 1934 that this concept was legislated by Congress, but the idea that the person verifying an organization's financial information should be independent of the organization being audited has for the past century been considered of paramount importance by the profession.

Even the U.S. Supreme Court has weighed in on the subject. In 1984, in United States versus Arthur Young & Co., the Supreme Court concluded that the independent auditor owes ultimate allegiance to the corporation's creditors and stockholders as well as to the investing public. This function, said the court, demands that the accountant maintain total independence from the client at all times and requires complete fidelity to the public trust.

Although "independence" is fundamentally a state of mind, not a designation that can be determined by a specific litmus test, the concept was nevertheless redefined and strengthened many times over the years by both federal and state legislation, and by the self-regulatory efforts of the accounting profession. (See Chapter Four.)

The theory is simple. The capital system as currently constituted in the United States is based on trust. In order for the system to work properly, the people who use the financial information being disseminated by public companies and other institutions must have faith that what is being reported is reasonably accurate. But if investors or other interested parties are to have confidence in financial information, it should be reviewed by someone not beholden to that organization—in other words, someone without an ax to grind.

The *Report of Independent Public Accountants* also states that an audit's objective is "to obtain reasonable assurance that the financial statements are free of material misstatement." The CPA, licensed to audit these financial statements, is placed in a unique position between those who are communicating financial information, on the one hand, and those who will be using that information, on the other. Like a referee, it is the CPA who declares that the information being communicated is reasonably reliable—not absolutely accurate, but that it falls within reasonable bounds and that the information is being communicated in accordance with certain ground rules that have been established over the years. The auditor's opinion, in other words, provides users with a degree of "assurance" that the information they are receiving is reliable.

The auditor has a responsibility for determining that an organization's financial statements are not "materially misleading." But the word "material" has been defined any number of ways. The legal definition, taken from a second circuit federal court decision, is whether or not the information would make a difference to a reasonable person in making a financial decision.

It used to be that while an audit might well find fraud if it had occurred, the auditor had no responsibility to look for it. But the auditor's obligation to look for and find fraud that could result in material misstatements of the financial statements increased enormously between 1980 and 1995, and today there is much more of a positive obligation to search for it. In fact, since 1988 an auditor is indeed obligated to "design the audit to provide reasonable assurance of detecting errors and irregularities."

Not all companies need full-blown audits. Private companies, for example, have no legal obligation to comply with the securities laws. Consequently, some smaller companies frequently ask CPAs to "review" their financial statements. This involves the CPA making comparisons and asking questions to see whether the information makes sense, but without detailed verification. Others may simply ask CPAs to "compile" information in the company's books and record it in financial statement format. In many cases, such "reviews" or "compilations" are sufficient for the needs of bankers and other grantors of credit.

EXPECTATION GAP

Much of the effort of the Treadway commission, as well as the profession's many other committees and studies aimed at preventing and detecting fraudulent financial reporting, was directed toward correcting the widely reported "expectation gap." This was a problem that had been identified during the previous decade, and targeted in particular by the Institute's Commission on Auditor's Responsibilities (the Cohen commission, named after its chairman, Manuel Cohen). It refers to the gap between what the public expects of the accounting profession, particularly as it relates to the audit function, and what the profession understands is its proper role.

The problem has been discussed and studied for years, and solutions have generally centered around ways to improve the communications between auditors and readers of financial statements. At the October 1985 Council meeting, for example, Jerry Sullivan, then incoming chairman of the ASB, told us he believed there were four major areas in which standards setters could help narrow the expectation gap:

1. Clarify the auditor's responsibility for evaluating internal control systems and detecting and reporting fraud.
2. Improve the communication of audit conclusions to annual report readers.

3. Improve the communications to financial statement readers of the business risks undertaken by management in the daily conduct of their business. (Rather than a mere transcript of historical events, readers want forward-looking information that will help them predict future returns.)

4. Assist auditors in making the key judgments and evaluating the key estimates critical in the preparation of financial statements. (For example, more practical and timely guidance was probably needed to help auditors assess the reasonableness of bank and insurance company reserves.)

Numerous public opinion studies during these years showed that the general public, including even sophisticated users of financial information, believed it was the auditor's responsibility to detect fraud. One of the most comprehensive studies, conducted in 1986 for the AICPA by the Louis Harris polling organization, surveyed both the public and "leaders," which included businesspeople, bankers, attorneys, legislators, security analysts, and members of the financial media. The study found that two of every five (39 percent) leaders—most of whom regularly used financial information in the performance of their employment—believed that an auditing firm's unqualified or "clean" opinion regarding a company's financial statements meant that the firm was verifying that all figures were completely accurate. Well over half (56 percent) of the general public agreed.

The profession, on the other hand, had over the years been reluctant to give any such assurances. Moreover, the fear of litigation created a conflict between what auditors believed in their hearts was their responsibility and what they were willing to say publicly.

This dichotomy hasn't always existed. At the beginning of the twentieth century, fraud was equated with theft of assets and detection was at the heart of the audit. Then came the stock market crash, the creation of the SEC, and a greater emphasis on the presentation and disclosure of financial information. The term "fraud" came to be more closely identified with intentional misstatements of financial information. Subsequently, the Institute's official auditing literature moved to the view that the ordinary audit wasn't designed to search for either type of fraud, even though it might be found during the audit process.

But that proved to be unworkable. Were we or were we not responsible for detecting fraud? "Yes and no" was not the answer people wanted to hear. The courts and everyone else believed auditors were responsible for detecting fraud, and no amount of explanations on the part of the profession was going to convince them otherwise. The problem was only compounded once the auditing literature began to move to the view that we weren't responsible, since that only gave many auditors the excuse to stop looking for it. So the literature was again changed to explain that the auditors should plan their work to look for fraud, while at the same time recognizing that under some circumstances—particularly when management conspired to deceive—it might not be found. To many, and unfortunately for the profession to many juries, this still sounded like doublespeak.

Many of us at the AICPA recognized that the profession had to take it on ourselves to close this expectation gap, not by moving the public perceptions to ours, but quite the opposite. We had spent decades trying to convince our constituencies that we weren't responsible for finding fraud, but the Louis Harris poll had shown us unequivocally that this strategy hadn't worked. Instead, we needed to become more vigilant in detecting fraud in an audit, but we also needed to make certain that the other Treadway commission recommendations aimed at sharing this responsibility with the corporations themselves were implemented.

We knew we could never guarantee that an audit would always detect fraudulent financial information, particularly if it had been purposely provided to the auditors by corrupt corporate management. But we also believed that there were many steps that could be taken to increase everyone's vigilance, and to make all parties understand that this had to be a shared responsibility.

The findings of the Treadway commission, supported by other studies conducted by the profession and by outside regulatory agencies, led the ASB in April 1988 to issue a series of nine new Statements on Auditing Standards (SAS). Collectively, they were referred to as expectation gap standards because they were designed to close the expectation gap.

The new standards represented the first changes since 1948 in the standard auditor's report found in every public company's annual report. For the first time, accountants began telling shareholders that they have planned their audit to provide "reasonable assurance" that the financial records were "free of material misstatement."

We were finally embracing, as much as possible, the responsibility that the public had been bestowing on us all along. For the first time, these new guidelines made it clear that auditors had a responsibility to find financial fraud, rather than simply detecting it if it happened to come their way.

In another attempt to close the "expectation gap," a sentence was also added to the auditor's report making it clear that the financial statements had been assembled by management. The Institute's research had long revealed that investors typically held broad misconceptions about who prepares financial statements. Many people felt they were prepared by outside accountants or even by the SEC. Similarly, because many investors incorrectly believed that accountants review every transaction, particularly those detailed in the annual report, a clause was added stating that accountants review documents on a "test basis" only.

The new SAS also required auditors to inform the company's audit committee about any irregularities that had been detected, and to recognize that under certain circumstances the auditor had a duty to report such irregularities to an outside authority. Auditors also now had to disclose any doubts they might have that the company they were auditing could survive another year.

The new expectation gap standards were certainly helpful, and were part of an ongoing, comprehensive effort to demonstrate that the profession was improving its efforts to become more effective in detecting fraud. The expectation gap by no means disappeared, however. In fact, in 1993, a public opinion survey by the Peter Hart organization revealed that although auditors now had an affirmative responsibility to detect fraud, the expectation gap between auditors and users had apparently not narrowed. Perhaps this was because while the public expected auditors to detect "all material financial statement fraud," even the new standards placed limitations on the auditor's responsibility. For example, auditors who performed their work diligently were not held accountable when there was collusion or forgery, and there were still substantial limitations on the auditor's obligation to disclose fraud to the investing public.

The Peter Hart study was commissioned by the National Accountant's Coalition, established in 1993 by the Big Six to investigate the liability crisis for themselves. It revealed that 70 percent of even Big Six clients, ostensibly the most sophisticated users and disseminators of financial information, believed that an unqualified audit guaranteed the accuracy of the financial statements. "Peter Hart's advice to us was that we've lost the public relations battle, that we should forget about trying to qualify our responsibility to detect fraud and just accept what our clients were telling us about their expectations," recalls Lee Blazey, the coalition's executive director.

Given the continued expectation gap, it was no wonder that even after these important changes were made concerning an auditor's responsibility to detect fraud and irregularities, the ASB continued to reinforce our responsibilities with incremental improvements. In late 1993, for example, it began debating new standards to further strengthen the guidelines for auditors to assess the risk of fraud when conducting an audit. Finalized three years later, they required auditors to look for a lengthy list of risk factors that in previous cases of fraud had been found to exist.

While the 1988 rules were important because they were the first to require auditors to search actively for fraud, they were rather vague about how auditors should go about it. Under the new rules, auditors now had to document precisely how risks should be assessed. Auditors were also now required to step back and consider whether a company's overall behavior demonstrated a potential for fraud, rather than just look at individual indicators such as receivables or inventories. If, for example, managers were being pushed to achieve unrealistic earnings gains, or if a company reported unusually large year-end sales, alarms should go off. Specifically, these red flags were defined as:

- Excessive management interest in boosting the stock price or earnings results through aggressive accounting
- Unduly ambitious financial targets set by management for operating personnel
- Inability to generate cash flow from operations while reporting earnings growth
- A management compensation plan that depended on meeting aggressive operating or cash flow gains
- Bank accounts or subsidiaries in tax haven countries, without a clear business reason
- Significant or unusual transactions, particularly if they occurred close to the end of the year

A POLITICAL SOLUTION

While the profession was doing everything it could during these years to improve the audit process, it was becoming increasingly clear to us that the solution to our liability crisis, if there was one, lay in more effectively voicing our concerns to state and federal legislatures. In fact, as we shall see in the following chapter, the nation's capital became a focal point for a host of activities that would profoundly affect our profession.

Chapter 3

Coming of Political Age:
Fifty-Four Jurisdictions, the SEC,
and the U.S. Congress

Medical doctors, and perhaps even attorneys, might disagree, but it seemed to me during my fifteen years as AICPA president that accountants had more layers of regulation than any other profession. Congress, the courts, dozens of state and federal regulatory agencies, several independent bodies, and the accounting profession itself were all intent on prescribing our rules.

To begin with, each of the fifty states, as well as the District of Columbia, Guam, Puerto Rico, and the U.S. Virgin Islands, is responsible for licensing, regulating, and disciplining the CPAs who practice in their individual jurisdictions. Auditors of public companies are then regulated by the federal government, primarily through the SEC's oversight of the financial reporting process and the Internal Revenue Service's monitoring of the tax code. Auditors of any organization that receives government funds also must follow the regulations prescribed by the funding agency.

Not to be left out, the accounting profession, under the auspices of the AICPA and the individual state CPA societies, developed some stringent regulatory programs of its own. In fact, between 1980 and 1995 self-regulatory activities, led by peer review and our professional code of ethics, increased at a greater rate than government regulation.

There are also a number of independent bodies that are not beholden to any branch of the state or federal government, or to the accounting profession. The Public Oversight Board (POB) and the Financial Accounting Standards Board (FASB) are two such organizations. And finally, every strand in this intricate web of regulation operates under the watchful eye of the U.S. Congress, the SEC, and the judicial system.

STATE SOCIETIES AND ACCOUNTANCY BOARDS

With all the hoopla that often surrounds federal regulation of the accounting profession, the amount of state regulation is often overlooked. Remember that there is no national certification of CPAs. We are certified and licensed only by the boards of accountancy in each of the fifty states and four federal jurisdictions (Guam, the U.S. Virgin Islands, Puerto Rico, and Washington, D.C.). Established by state legislation, these boards have full authority to regulate the practice of public accountancy within their jurisdictions. Each such board has its own guidelines of professional conduct and can take disciplinary action against licensees who violate their rules. They alone have the authority to revoke, suspend, assess, fine, or otherwise regulate a CPA's license to practice. They also typically have the authority to insist that an individual accountant take specific remedial action, including additional CPE requirements and follow-up peer reviews.

The state societies also have a powerful presence in the AICPA's Governing Council, since Council positions are allocated to the states in relationship to their population. Consequently, the largest half-dozen or so states could constitute a majority if they voted as a block.

One of my first priorities as AICPA president was to improve the relationship between the Institute and the state societies, which historically had all too often been shut out of the AICPA's planning process. That didn't make sense. Our efforts to better our profession needed to be a partnership. After all, about 80 percent of AICPA members also belonged to their individual state associations. We were serving the identical membership. During the early 1980s I made it a point to meet with as many state society presidents and executive directors as possible to emphasize the idea that we were all in this together.

UNIFORM ACCOUNTANCY ACT

State regulation of CPAs is no doubt here to stay, but the lack of a national certification of CPAs is actually an anachronism—more an accident of history than anything else. Unlike the legal profession, where laws often differ among the states, the practice of accountancy is national in scope. Financial reporting and auditing standards don't vary. States may have different tax laws, but they are easily learned.

But while accounting and auditing standards are uniform in all fifty-four jurisdictions, licensing requirements remain a mosaic of different rules and requirements, making it difficult for CPAs to move from one jurisdiction to another. In fact, a handful of states, called "title states," don't even restrict the audit function to CPAs. They legislate only who can call themselves CPAs, but allow anyone to perform an audit.

We did, however, make great strides between 1980 and 1995 toward adopting a degree of uniformity concerning certification requirements. The Institute tried to take the lead in this regard by urging the adoption of model legislation jointly drafted by the AICPA and the National Association of State Boards of Accountancy (NASBA). We also raised AICPA membership requirements and hoped that state accountancy boards would follow suit. Almost all jurisdictions, for example, adopted our membership requirements of 150 college-level semester hours of education and CPE (see Chapter Five).

But the road to uniformity during these years was not an easy one. By 1980, there had been at least a dozen versions of a model bill. The first one had been drafted in 1916 by the American Institute of Accountants, the predecessor of the AICPA, and the last one in 1981. But in April 1980, NASBA published its own Model Accountancy Act. By promoting two different bills simultaneously, the profession was sending a confused message to state legislatures. Even when there had been one bill, states chose the sections they liked and ignored the rest. Two bills complicated matters further, since there were significant differences between the two ap-

Table 3.1. Differences between AICPA and NASBA Bills

AICPA bill	NASBA bill
Required a reciprocal certificate as a prerequisite to obtaining a permit to practice in another state	Only one CPA certificate required, valid in other states
Two years of experience	One year of experience
Peer reviews of CPA firms as a condition for renewal of a firm's permits to practice	No policy on peer review
State boards permitted to levy fines as a disciplinary action	Disciplinary actions instituted by the state's attorney general.
Provided for a special examination as an alternative to fulfilling CPE requirement.	Allowed no such trade-off

proaches, including requirements concerning reciprocity, experience, peer review, CPE, and disciplinary proceedings (see Table 3.1).

It was important that we develop a unified approach, so in 1983 the Institute participated in the AICPA/NASBA special committee on the model accountancy bill. Our representative was AICPA board chairman George Anderson. The other two members were former NASBA president Leighton Platt and Colorado state society president Gordon Scheer. The purpose of the committee was to produce a model bill that emphasized those provisions on which the AICPA and NASBA agreed.

By October 1984 we had narrowed our differences to the experience and peer review requirements. The bill was subsequently published with a commentary explaining both sides of those two issues, while emphasizing our common ground.

But having a bill that offered alternatives on even these two issues didn't exactly provide for a cohesive message. So perhaps it shouldn't have been much of a surprise that we still didn't have much success in persuading state legislatures to pass the bill. It seemed like an oxymoron to have a joint bill with differences.

It took us until 1992 to reach a compromise with NASBA, but it is a credit to the men and women who worked so hard on this effort that it was accomplished at all. The newly titled Uniform Accountancy Act (UAA) mandated that all CPAs must meet certain CPE requirements. It also met NASBA's concern that the requirement was too stringent by directing state boards to acknowledge the value of courses related to untraditional accounting services.

We also reached a compromise on the experience and peer review requirements. The Institute's suggestion was to eliminate the experience requirement altogether. (With the 150-hour education requirement and the uniform examination, we believed it was unnecessary.) But NASBA made it clear that many of their state boards would not go along with that. In fact, many of them wanted two years, so we compromised on one year of experience in the practice of public accountancy or its equivalent under the supervision of a licensed CPA. We also agreed on mandatory quality review every three years. To satisfy NASBA on that issue, the UAA called for a state quality review oversight body to report to state boards on the effectiveness of quality review programs.

From the AICPA's point of view, the important thing was that the key ingredients for entry into the profession under the UAA were the same as they had been under the many model accountancy bills that had come before it, namely, the "three Es"—education, examination, and experience. It mandated that CPAs had to have a minimum of 150 semester hours of college credit, must pass the uniform CPA examination, and must have at least 2000 hours of work experience. The UAA also mandates 120 hours of CPE during the three-year period preceding renewal of one's CPA license, with a minimum of 20 hours each year.

CPA OWNERSHIP OF ACCOUNTING FIRMS

Until recently, the UAA also stipulated that CPAs could offer accounting services only through a CPA firm. But that rule, which had also been an AICPA membership requirement, was successfully challenged in the courts, most notably in a series of cases in Florida and Texas. As a result, the UAA allowed CPAs to offer nonattest services through any type of entity they chose, whether or not it was owned by CPAs. The AICPA rule that attest services could be offered only through a firm controlled by a majority of CPAs was included in the UAA.

This change was the subject of some rather intense debate over the years. At the direction of the Council, in March 1993 we established a Task Force on Non-CPA Ownership, chaired by Ronald Cohen. After modifications to respond to suggestions made at various regional Council meetings, Ron offered a resolution at the May 1993 meeting proposing non-CPA ownership for active employees of accounting firms, but opposing any outside ownership. It would then be up to each state to decide for itself if its rules should change in accordance with the AICPA recommendations.

"But at least the AICPA would provide an outline for consistency and will not be standing in the way of states making these changes," Ron explained.

While the motion allowed for non-CPA ownership, it still mandated some strict requirements, including:

- A majority of the ownership of the firm in terms of financial interests and voting rights must belong to CPAs.
- Non-CPA owners must be actively engaged in the firm as his or her principal occupation in providing services to clients.
- A CPA must have ultimate responsibility of all work performed by the firm.
- Non-CPAs becoming owners after adoption of the Council's resolution must possess a baccalaureate degree and, beginning in the year 2010, have obtained 150 semester hours of college education.

- Non-CPA owners must complete the same CPE requirements as AICPA members.

Jack Kennedy, eleven years on the Council, a board member, and former president of the Pennsylvania Society, spoke in favor of the resolution, asking rhetorically, "Isn't it really an issue between heart and head? Heart is about from whence we came, the good old days, single-focused practices, a less complex society. Head is about growth, about a diverse practice, about recognizing the reality of change. While I appreciate our hearts, I strongly urge that you vote with your head."

But Norm Lipshie from New York State pleaded, "Maybe we should go back to the heart. Maybe we should go back to the definition of a profession as a body of specialized knowledge, a formal education process, standards governing admission, a code of ethics, our recognition of a public obligation. That's a profession."

As a compromise, Joseph Kenyon, a Council member from Minnesota, offered an amendment to require that a super majority of two-thirds, rather than a simple majority, of ownership of accounting firms had to belong to CPAs. That amendment narrowly passed, 117 to 102, after which the main motion passed as well, 155 to 58.

As will be demonstrated shortly, this was by no means the first time pragmatism had won out over idealism. "Basically what this resolution is about is recognizing change," noted Curt Mingle, a Council member from Illinois and a member of the AICPA board of directors. "We can't put our head in the sand."

SUBSTANTIAL EQUIVALENCY

Another important goal of the UAA was to make it easier for CPAs to practice across state lines, both in person and electronically over the Internet. That's why the Act includes a "substantial equivalency" provision, which means that any CPA licensed in a state that has "substantially" adopted the certification criteria of the UAA is qualified to practice in any state that has also adopted the most important elements of the UAA. Each state board of accountancy determines for itself whether another state's certification requirements are "substantially equivalent" to their own, but it can also request a determination from a newly created NASBA Qualification Appraisal Service. The Act also contains specific provisions for the licensing of Canadian chartered accountants under the substantial equivalency concept.

One of the profession's most ardent proponents of substantial equivalency during these years was Bob Mednick, a senior partner with Arthur Anderson who became AICPA chairman of the board in 1996. He liked to point out that most CPAs who operate across state borders are in technical violation of accountancy rules. That's because if a client in New York moves to Florida and still wants his accountant to do his taxes, the New York CPA is likely unaware that he or she is theoretically supposed to be licensed also in Florida. The problem became even more pronounced as the Internet made interstate practice so effortless.

While Bob and others were optimistic that practicing in multiple states simultaneously would someday require no more paperwork than driving coast to coast, I frankly think it will be many years before full reciprocity among most states is achieved. In the first place, states aren't about to give up their sovereignty on this issue. It also doesn't seem to be a major priority on the part of most CPAs. While certainly there are many accountants with national and international practices for whom a lack of reciprocity is a serious impediment, the fact is that the vast majority of accountants have very local practices.

"The trend is clearly moving toward more and more CPAs having a national or international practice, and the Internet will precipitate this," explains Jim Kurtz, former executive director of the California Society of CPAs. "But the degree is more anticipated than urgent. Even in California, I do not yet see a critical mass of national, much less international, commerce among the majority of our members to support the concept of making a CPA license as transferable between states as a driver's license. When the marketplace demands it, it will happen, but probably not before."

A NATIONAL CPA CERTIFICATE

Jim was expressing an important sentiment. The overall goal of the UAA was to protect the integrity of the CPA designation by seeking to make the regulations in all fifty-four jurisdictions meet certain standards. Every jurisdiction, for example, had made passing the uniform CPA examination a condition for qualifying for a CPA certificate. (See Chapter Five.)

Still, complete compliance to the UAA remains a distant goal, mainly because there continued to be so much disparity among the states concerning education and experience requirements. For example, California has not adopted the 150-hour education requirement, and apparently had no plan to do so. Instead, it has a version of what is known as an "Abraham Lincoln provision," whereby access to a professional examination is open to anyone by means of a waiver process. All those who can pass the examination and show the required experience can be licensed as a lawyer, doctor, or accountant. That makes it difficult to increase the educational requirements for these professions.

There is also a hodgepodge of different experience requirements, with some states having no requirements at all, while others requiring a very specific number of hours in each of several areas of expertise. Even those jurisdictions with an experience provision typically only require a letter from a CPA certifying that the applicant has been employed for the requisite period of time.

Frankly, many of us became increasingly frustrated over our inability to come up with a consensus among the states. Eventually I had to admit, to myself at least, that we would not be creating order out of this chaos any time soon. In fact, given the difficulty of fifty-four jurisdictions agreeing on anything, much less detailed rules for certifying accountants, I began to wonder if another solution might be more effective. Rather than more regulation at the state level, what about less regulation, or deregulation? If states began forfeiting their certification responsibilities, the AICPA membership requirements would likely become the de facto standard.

This is just what the AICPA's committee on regulation, affectionately known as the "doomsday" committee, was charged with studying during the early 1980s. It was a difficult period for states financially, and there was a nationwide movement under way to trim state government spending and to reduce the bureaucracy by paring government agencies. Many states enacted sunset laws, requiring that every five years an agency must rejustify its existence to continue. Many state CPA agencies could not pass the test because they couldn't point to any instance in their history in which they had either suspended or revoked the license of a CPA.

The doomsday committee, chaired by my former partner, Leroy Layton, was appointed to consider quietly the question of what should be done if an individual state did away with its licensing statute altogether. At the time, this was a real concern in several states. The committee's final report suggested that if one or more states eliminated its CPA statute, the Institute, in cooperation with the state CPA society, could create a national CPA credential in that state.

I was intrigued with the idea, believing that we might have a better chance of achieving a national standard if we didn't need the support of every state legislature—an almost impossible task. I felt that strong arguments could be made to states that they could rely on the profession for regulation rather than using limited state resources.

Unfortunately, I could never get anyone else among the AICPA leadership behind the idea of a national certification program. Mainly, I think the inertia was based on a preference for the devil we knew over the devil we did not. The concern was that if we dropped state legislation as the mechanism for licensing CPAs, we would be giving the federal government the opportunity to step in and control the profession. Once we started down that path, we had no assurances that the process could be controlled. I certainly understood that argument.

As a consequence, the profession continues to support licensing of CPAs at the state level. But personally I think the jury is still out on whether we will ever be able to successfully persuade all fifty-four jurisdictions to agree on a single set of certification requirements.

SEC ON THE ATTACK

While the fifty-four states and territories are solely responsible for giving CPAs their license to practice, most of the regulatory action between 1980 and 1995 occurred in the nation's capital. As already discussed, my tenure as AICPA president began in the midst of a major litigation crisis, with all the large accounting firms (and many small ones) paying the price for a handful of high-profile business failures. But it wasn't just the courts holding us responsible. The SEC, as well as the U.S. Congress, was also putting our collective feet to the flame, and we felt its heat.

Several high-profile bankruptcies occurred in 1983, during which auditors were criticized for not detecting fraud or emerging financial problems before they caused a bank failure or a corporate earnings plunge. Whether it was Peat Marwick giving assurances of Penn Square Bank's good financial health just months before bad energy loans caused it to fail, or Ernst & Whinney giving an unqualified audit to the United American Bank in Tennessee thirty days prior to its collapse, the profession felt the pressure from what seemed like almost daily newspaper accounts implying that the accounting profession was to blame. It was an uncomfortable environment for us all.

Responding to these kinds of well-publicized reports, the SEC, too, went on the attack. In 1983, in quick succession it:

- accused the accounting firm, Fox & Company, of failing to spot misleading data in the financial statements of Saxon Industries, which later filed for bankruptcy
- was responsible for the indictment of an Arthur Andersen senior partner for issuing a false financial statement about the bankrupt Drysdale Government Securities
- charged a Coopers & Lybrand partner with conducting a deficient 1979 audit of the Security America Corporation, which subsequently was forced into liquidation
- censured Touche Ross for failing to use GAAS in its examination of the financial statements of Litton Industries

FIRMS QUESTIONED OVER LOANS

The SEC was becoming increasingly vigilant in going after accounting firms that in its view had not maintained their independence from clients. During the S&L crisis, it was particularly concerned with auditors who had been given personal or business loans by the banking institutions they were auditing.

In 1991, the SEC filed suit against Ernst & Young, accusing it of acting improperly by auditing the books of Republic Bank Corp. of Dallas, Texas, at the same time that more than fifty of its partners had about $16 million in loans from the bank. The SEC charged the accounting firm's predecessor, Arthur Young & Co., with failing to disclose that the loans had compromised its status as an independent auditor. According to the SEC, at least twenty-seven partners had received personal loans from the bank. The complaint further charged that many of the loans were unsecured and were not made under normal lending procedures, terms, or requirements. Making matters worse, of course, was that in 1987 the bank failed.

These accounting firms were at a disadvantage because at the time, the rules concerning auditors receiving loans from banks they were auditing were fairly ambiguous. Partners were permitted to borrow money from clients as long as the loan was secured (such as a home mortgage), was immaterial to a partner's net worth, and was on normal and reasonable terms. But these rather vague criteria were open to confusion and abuse. As a result, the AICPA changed its rule so that accountants were forbidden from obtaining any new loans from clients, and were allowed twelve months to refinance loans from institutions that became clients.

CONGRESSIONAL OVERSIGHT

While the SEC has primary regulatory responsibility over the accounting profession, congressional oversight has always been of great concern because of Congress's ability to legislate, and thereby regulate, all aspects of accounting. Congressional scrutiny predated, of course, my appointment as AICPA president. In fact, during the preceding decade a Senate Subcommittee of the Government Affairs Committee, chaired by Senator Lee Metcalf of Montana, held extensive hearings to investigate why failures and wrongdoing by publicly owned corporations had not been detected or disclosed by independent auditors.

These 1976–77 hearings were, indirectly at least, an outgrowth of the Watergate scandal, which had revealed that some of the country's largest corporations had almost routinely been making illegal political contributions. That led to the disclosure that at least 200 of them had also been using secret accounts to make illegal payments, particularly unrecorded foreign bribes and kickbacks. After extensive investigation, Congress enacted the Foreign Corrupt Practices Act of 1977, which outlawed a variety of foreign payments. The act also required SEC registrants to maintain internal accounting controls to provide reasonable assurance that all transactions were authorized by top management and recorded in conformity with generally accepted accounting principles.

Within this environment, Congress then set out to make a broader investigation of the accounting profession. Throughout 1976, the Senate subcommittee was concerned primarily with the same two issues Congress would revisit a decade later, namely, improving the quality of audits, and auditor independence.

Early in 1977, the subcommittee issued a 1760-page staff report critical of the profession—so critical in fact that my predecessor, Wally Olson, called it "almost as damaging to the profession as the Japanese attack on Pearl Harbor was to the U.S. Navy in 1941."

This was no doubt hyperbole, but the report's sixteen recommendations for reform did boil down to a single strategy of reversing more than four decades of policy by taking away the accounting profession's self-regulatory responsibilities and transferring them to the federal government. The suggestion that the SEC or some other federal agency should establish rules for audits of publicly owned companies and enforce standards of conduct and periodic review of auditors was particularly troubling to the AICPA leadership, who cared deeply about maintaining the independence, professionalism, and public responsibility of accountants. The staff report's other onerous recommendations included the suggestion that the fifteen largest accounting firms should be required to file annual financial statements with the SEC, and that both the SEC and the Federal Trade Commission (FTC) should investigate alleged antitrust violations within the profession.

The Institute had no choice but to take these threats seriously, particularly after Congressman John E. Moss, Democrat from California, who chaired hearings in the House of Representatives comparable to the Metcalf hearings in the Senate, introduced legislation that many observers believed had a better than fair chance of passage. It would have established a new federal agency, the National Organization of Securities and Exchange Commission Accountancy (NOSECA), to regulate, control, and discipline all CPAs who audit public companies.

The AICPA immediately mobilized to convince Congress that such drastic steps were not needed. We believed strongly that in order to maintain the confidence of public and institutional investors, the accounting profession must maintain its independence from industry, and from government. Under Wally Olson's leadership, responses were prepared, our lobbying efforts in Washington were intensified, and the profession slowly began to acknowledge that it would have to strengthen its own self-regulatory efforts if it were to forestall legislative action. We also began to recognize the growing importance of having a more effective operation in Washington.

Fortunately, however, as the 1970s drew to a close, congressional pressure for reform of the accounting profession began to dissipate. I think it's fair to say that this occurred more as a result of circumstances beyond our control than anything the profession did to forestall it. First, Senator Metcalf, the profession's leading critic in the Senate, died in December 1977, and Congressman Moss, the House's leading critic, did not seek reelection. And second, the Washington establishment was distracted by other, more pressing matters, particularly President's Carter's legislative agenda for coping with the oil crisis.

DINGELL HEARINGS

Perhaps the fact that none of the major issues investigated during the 1977–78 Moss–Metcalf hearings were truly resolved during those years made the congressional scrutiny that occurred eight years later all the more intense. On February 20, 1985, the House Energy and Commerce Subcommittee on Oversight and Investigations, chaired by Congressman John D. Dingell, Democrat from Michigan, once again began hearings to scrutinize the effectiveness and integrity of the accounting profession. In particular, it set out to determine whether the SEC had delegated too much of its oversight responsibilities to self-regulatory bodies, namely, the AICPA.

From the very outset, it was clear we were in for some stormy moments. Congressman Dingell opened the hearings by calling the accounting profession's attempts at self-regulation "weak," and warning that the profession faced "what may be its last opportunity to regulate itself instead of having somebody else do it for them."

"The system doesn't seem to be working," Dingell insisted. "If we absolve accountants from blame, perhaps we should absolve ourselves of the need for accountants and save ourselves a lot of money. If they are providing a useful service, it should be one on which the public can rely."

Several other Democratic subcommittee members were equally critical of the SEC. Congressman Ron Wyden of Oregon charged that it had "essentially defaulted its oversight responsibility to self-regulatory bodies and disciplinary action to private litigation."

These were strong words, and they certainly made us at the AICPA sit up and take notice. Many subcommittee members were particularly concerned about several spectacular cases of business failures that had occurred shortly after clean opinions had been issued by independent auditors. Particular attention was given to the collapse of such financial institutions as the Penn Square Bank, the United American Bank, Drysdale Government Securities, and Baldwin-United Corporation. Critics of the profession liked to call these disasters "audit failures," as if companies filed for bankruptcy and shareholders suffered monetary losses solely as a result of an auditor's failure to discover financial problems. As the congressional hearings began in the winter of 1985, we felt strongly that the accounting profession's share of responsibility for the bankruptcies of the 1970s and 1980s was relatively small. We were also fully aware, of course, that we had our work cut out for us if we hoped to convince anyone else of this.

We were intent on getting our point across that no one—not accountants, regulators, or Congress—had anticipated the oil crisis, the S&L debacle, or the recessions the U.S. economy had experienced during the previous three decades. Surely there was enough blame to go around for the more than 350,000 business failures that occurred between 1982 and 1986.

The S&L crisis, for, example, which forced the Big Six accounting firms to pay more than $1.6 billion in damages and settlements to investors, was in fact caused by a host of players and circumstances all coming together to create a system that wasn't sufficiently vigilant. By the early 1980s, problems in the thrift industry, particularly those in the Southwest, were well known. Rising interest rates had led banking regulators to allow institutions to expand into different and frequently riskier areas, such as commercial real estate and business and consumer loans. Also contributing to an increased number of uncollectable loans was a sustained drop in oil and real estate prices.

Deregulatory efforts to help troubled institutions continue to operate in the hope that economic improvements would solve their problems proved disastrous. Regulators only made matters worse by allowing banking institutions to defer losses from the sale of assets with below market yields, to include mutual and income capital certificates in net worth, and to allow excessive up front income recognition for loan origination and commitment fees. Regulators also proposed that thrifts be allowed to defer and to amortize loan losses, and even favored withdrawing the requirement that they disclose potential losses in FSLIC-assisted mergers.

Yet rightly or wrongly, during the 1970s, and now again in the 1980s, the accounting profession was being held responsible—by many courts, but also by the U.S. Congress—when shareholders lost money, as they do after virtually every business failure. In particular, Congressman Dingell and several of his colleagues made no attempt to hide their belief that competitive pressures and the expansion of the major accounting firms into consulting were eroding the effectiveness and independence of auditors and that auditors had become too cozy with their clients.

AUDITOR INDEPENDENCE

At one point during the hearings Congressman Dingell held up a newspaper advertisement for the then Big Eight firm of Deloitte Haskins & Sells. The headline read, "Think of us as partners, Deloitte Haskins & Sells & You." The ad ended with the line, "In fact, there's only one thing wrong with calling yourselves, Deloitte Haskins & Sells & You. The 'You' really should come first."

"That doesn't sound too independent to me," Dingell pointed out.

A number of critics of the larger firm were also invited to testify. Robert Chatov, an associate professor of management at the State University of

New York at Buffalo, told the Subcommittee that the independence of auditors was being compromised because major accounting firms were increasingly providing partisan management consulting services to their audit clients. Abraham J. Briloff, an accounting professor at Baruch College, agreed and recommended that accounting firms be banned from rendering peripheral services to audit clients. And Eli Mason, chairman of the National Conference of CPA Practitioners, an alliance of small accounting firms, accused the Big Eight of aggressively cutting audit prices, some to below cost, in order to get a foot in a client's door.

It was within this environment that I was placed on the Subcommittee's hot seat on March 6, 1985. Fortunately, I had been well prepared by our Washington office. I began by explaining the common misconception that a "clean opinion" by an auditor on a company's financial statements is a guarantee that the company won't subsequently fail. In fact, I told the Subcommittee, such an opinion provides the public only with reasonable assurance that the representations of management reflected in the company's financial statements have complied with generally accepted accounting principles. My testimony also pointed out that while critics liked to ballyhoo a handful of spectacular business failures that occurred shortly after a clean opinion, alleged audit failures accounted for less than 1 percent of the 10,000 audits conducted each year on publicly held companies, and that there were only 18 SEC enforcement actions against auditors in 1984.

There were plenty of instances, I added, in which auditors *had* alerted investors to possible problems. Deloitte Haskins & Sells, for example, in March 1981, and then again in February 1982, had raised questions about the condition of Dallas-based Braniff International Corp., stating on both years' financial statements that in its opinion "there are conditions which indicated that the company may be unable to continue as a going concern." Braniff filed for Chapter 11 bankruptcy protection in May 1982. Likewise, in October 1980 Peat, Marwick raised questions about whether or not the Itel Corporation, a San Francisco-based leasing, computer, and transportation company, could continue as a "going concern." Three months later Itel filed for bankruptcy.

Much of my testimony before Congressman Dingell's subcommittee was spent documenting the self-regulatory changes that had been made by the profession since 1977, the last time the accounting profession had been investigated in depth by a congressional committee. But I actually spent a lot of my time on the stand being grilled about the minutiae of precisely what actions an auditor is obliged to take if he or she uncovers evidence of an illegal act. I tried to make clear that illegal acts on the part of a company's top management were extremely rare. I offered my assurances that the accounting profession was committed to the highest standards of excellence and integrity and was deserving of the continued confidence of the investing public and of Congress.

BROOKS COMMITTEE

Simultaneous with the Dingell hearings, the House Committee on Government Operations, chaired by Congressman Jack Brooks of Texas, was investigating the quality of government audits. Once again, its initial conclusions were rather devastating. Its October 1986 report entitled *Substandard CPA Audits of Federal Financial Assistance Funds: The Public Accounting Profession is Failing the Taxpayers* stated in part:

> Audits performed by certified public accounting firms play a vital role in efforts to assure accountability in the use of the more than $100 billion in Federal financial assistance provided to state and local governments each year. It is, therefore, absolutely essential that these audits be of high quality. Unfortunately, all too often they are not. The committee's recent review documents that an estimated 34 percent of CPA audits of Federal financial assistance funds fail to meet generally accepted government auditing standards.
>
> The public accounting profession is not fulfilling its responsibility to the taxpayers. Congress will not tolerate continued sloppy, unprofessional, substandard CPA audits of federal tax dollars.
>
> Dramatic improvements must be made in the quality of these audits. The public accounting profession, the General Accounting Office, and the Federal Inspectors General have all undertaken efforts to improve CPA audits of Federal financial assistance funds. While these initiatives appear promising and are to be commended, additional corrective actions are needed. Above all, the accounting profession must impose strict sanctions on CPAs who perform substandard audits and make these disciplinary actions public knowledge. Failure to take these steps could very well result in greater federal government regulation of the profession.

Congress was quite right to be concerned that the huge sums flowing from the federal government to the states were properly accounted for. The Government Operations Committee was intent on improving federal audit policies and procedures. It had played a leading role in the enactment of the

Single Audit Act of 1984, which required state and local governments receiving $100,000 or more per year in federal financial assistance to obtain an independent, organizationwide audit of their operations.

In making its recommendations, the Brooks committee stressed the importance of enforcing the Inspector General Act of 1978, which required inspectors general (IGs) to "take appropriate steps to assure that any work performed by non-federal auditors complies with the standards established by the Comptroller General." (Each regulatory agency of the federal government has an IG responsible for investigating and reporting on the effectiveness of organizational and administrative matters.) The committee then suggested that the inspectors general of all agencies that receive CPA audits of federal funds should strengthen their quality review systems.

The accounting profession was asked to take further action. The committee called on the AICPA, together with the state boards of accountancy, to impose strict sanctions on CPAs who performed substandard audits and make public the results of all cases in which disciplinary action is taken. The committee also urged that all CPAs auditing federal financial assistance funds be properly trained in government auditing, including CPE in government auditing and periodic peer reviews.

The accounting profession took these criticisms and recommendations very seriously. In fact, even before the 1986 Brooks committee report had been issued we were trying to respond to both sides of the problem. On the one hand, the federal government was criticizing CPAs for not performing quality audits of the recipients of federal funds. On the other hand, CPAs were complaining that government procurement policies were overly sensitive to price and didn't properly consider the technical qualifications of the CPAs who were hired. One way we were able to successfully mediate the issue was a colloquium held in Cherry Hill, New Jersey, in November 1980. Probably the meeting's most important achievement was the opening of effective communication lines between CPAs and government auditors.

TASK FORCE ON THE QUALITY OF AUDIT OF GOVERNMENTAL UNITS

A few years later we established a task force on the quality of audits of governmental units, which in March 1987 made twenty-five recommendations, grouped into categories it called the five Es:

1. Education of the auditor
2. Engagement of the auditor

3. Evaluation of the audit work
4. Enforcement of professionals standards
5. Exchange of information

As we looked at these twenty-five recommendations, we realized not all of them could be implemented by the AICPA. There were more than a dozen groups and organizations that had to be involved in the implementation, including the GAO, federal inspectors general, state auditors, state boards of accountancy and CPA societies, NASBA, and several committees within the AICPA. As a result, we helped form an implementation monitoring committee made up of representatives of fifteen organizations. Its task was to review implementation of each of the twenty-five recommendations. When implementation was not proceeding, the representative of the group responsible was held accountable.

This proved extremely effective, as twenty-two of the twenty-five original recommendations, as well as some additional ones, were implemented. (The three that were not implemented were inconsequential.) One of our most fundamental steps was to require auditors of government programs to complete CPE courses in the unique aspects of government accounting and auditing and to participate in peer review. As a result of these and other efforts, by Congressman Brooks's own admission by the end of the decade the AICPA had gone a long way toward restoring his confidence in the accounting profession's ability to conduct government audits effectively.

WASHINGTON OFFICE

As the onslaught from the SEC and Congress continued, it was clear that the pressures coming to bear on us had one thing in common. Almost all of them emanated from Washington, D.C. Even the liability crisis was due in part to legislative principles that worked to our disadvantage, such as joint and several liability and RICO, which to many of us was more a legislative abomination than a legislative principle. Yet to our growing frustration, the Institute's Washington operation was not up to the challenge of effectively representing our interests in these complicated times.

The first truly capable head of the Washington office was Gil Simonetti, who arrived there in 1977 but soon left the Institute to become Price Waterhouse's primary Washington liaison. He was followed by Ted Barreaux, who brought with him a wealth of experience from his former position in charge of congressional relations for the SEC and created a strong presence for the Institute in Washington.

But by 1985, as the federal government was impacting our profession like never before, we needed to create a more finely tuned single voice that could speak on behalf of the entire accounting profession. While the Institute had a knowledgeable and forceful Washington staff, each of the ten largest firms had also established its own independent representative in Washington. Collectively, they called themselves the "Tidewater group," because their first meeting had been held at the Tidewater Hotel in Maryland. But while the group occasionally managed to speak with one voice, more often they would lobby Congress individually. Only rarely were their activities coordinated with the Institute's Washington office.

"A dozen people were walking around the Hill speaking on behalf of the profession, each of them saying something different," recalls B.Z. Lee, former managing partner at Seidman & Seidman, who in 1985 was sent to Washington as an unpaid observer by AICPA Chairman Mike Cook.

At the same time, the largest state CPA societies had become so frustrated with being kept out of the loop regarding the Institute's activities in Washington that five of them were planning to open their own Washington office. That would have added to the mix yet another voice purportedly speaking for the profession.

Clearly something had to be done to revitalize our Washington operation, increase the Institute's political acumen, and coordinate all the profession's activities into one cohesive message. One important catalyst for action was a detailed memo issued by the Tidewater group detailing precisely the ineffectiveness and disorganization of our current efforts in Washington. As a result of these and other signals we were getting urging us to take action, in 1986 Mike Cook's successor, Marvin Strait, and I asked B.Z. Lee to become deputy chairman of the Institute and to move to Washington to take charge of all our activities there as head of the Federal Affairs Committee.

His first order of business was to eliminate the internal warfare. He created a new Government Affairs Task Force comprised of all the groups speaking for the accounting profession in Washington, including the Tidewater group and the state societies. The task force met monthly, and often invited others to participate as well, such as the chairs of the AICPA's tax division and the ASB.

B.Z. had his work cut out for himself with his efforts to encourage the state societies and the largest firms to cooperate with one another and with the Institute. One major stumbling block was that prior to his arrival, there was little cooperation between the largest eight firms that were intensely competing in the marketplace. "Before B.Z. arrived, if I told my firm I had just heard from an Arthur Anderson partner about some important action about to be taken on Capitol Hill they would have been dumbfounded," recalls John Hunnicutt, an original member of the Tidewater Group while a principal at Peat Marwick, and B.Z.'s successor as head of the Washington office in 1993. "That kind of cooperation was unheard of."

But gradually the new Washington staff was able to convince the CEOs of the largest accounting firms that working together on common legislative issues did not violate antitrust law, and was the only way we could build an effective presence in Washington. Soon the CEOs of the top firms began to meet regularly too, although always accompanied by their legal counsel to be certain that the only issues on the agenda were their common interests in Washington.

Under B.Z.'s leadership, the AICPA's Washington office was gradually turned into an effective legislative force. It was further buoyed by the addition of other key personnel. B.Z. brought in John Sharbaugh, a former state society executive director, to coordinate state society activities, Tom Higginbotham as vice president of political affairs, and Don Skadden, a highly regarded tax practitioner, as the Institute's first vice president of taxation.

"B.Z.'s leadership was a breath of fresh air," recalls Jim Kurtz, executive director of the California Society of CPAs. "Suddenly all our efforts were collaborative."

KEY PERSON PROGRAM AND PAC

B.Z. began making important changes in our Washington operation. All of us among the AICPA leadership quickly recognized the importance of maintaining an effective Washington office with substantial research capabilities and with the resources to influence federal legislative and regulatory activities. But we also recognized that there were some areas in which the local, grassroots state organizations could do a much better job than the national Institute. For that reason, our Washington office began to more strongly encourage individual members to maintain relationships with their local members of Congress.

As a result, the AICPA developed a database of more than 1500 key persons who were constantly kept informed about the Institute's positions on current legislation. They were also alerted whenever it would be helpful for them to put in a call to their local Congressperson. All this was accomplished in a collaborative effort with the state societies, each of which had a key person coordinator.

The AICPA's key person program was developed as a partnership between the AICPA and the state societies. It helps the state societies focus on the most important issues affecting their members, and recognize just how much all of us have in common. The state societies, in turn, conduct their own lobbying of state legislatures, although particularly the smaller states often ask the Institute's Washington office for help in researching and analyzing complex issues.

Another important change we made in Washington during these years was to increase substantially the amount of money our profession's Political Action Committee (PAC) dispensed to political candidates. Equally important, we changed the way it was distributed. Prior to B.Z.'s arrival in Washington, the money was distributed centrally by the Washington office. One of the most important ways we made the key person program effective is that we began to funnel our PAC money through the individual states to their local politicians. It helped, of course, that we began to have considerable sums to distribute. In 1980, our PAC contributed only $30,000 to political candidates, in 1982 less than $50,000, and in 1983 less than $20,000. But in 1987 we added a voluntary contribution check-off on the AICPA membership dues form, and as a result raised nearly $1 million that went directly into our PAC. Suddenly we had one of the largest PACs in the nation, and our political clout grew accordingly. By 1994, the final election cycle of my term as AICPA president, our PAC dispensed almost $2 million to political candidates, 80 percent of which was handed out at the local level. (The Institute would welcome efforts to reform the way political campaigns are funded, but in the meantime this is, unfortunately, the way the political game is played.)

By the time B.Z. turned the reigns over to John Hunnicutt in 1993, our Washington operation was organized into four divisions:

1. A tax division
2. A federal government division, which was our nexus to federal regulatory agencies
3. A congressional and political affairs division, which included our team of lobbyists interacting with Capitol Hill and managing our PAC

4. A state legislation committee, which concentrated on supporting the state societies

One seemingly innocuous, but ultimately extremely successful, activity begun during these years was the AICPA's Digest of Washington Issues. This recitation of the issues affecting the accounting profession, along with our concerns and proposed remedies, initially was intended for the leadership of the profession, but soon it began to be distributed widely to members of Congress and other professionals on Capitol Hill. Many of these people in a position of influence began to rely on it as they made their legislative decisions. It also was distributed among our membership, which resulted in a much better informed profession.

The combination of our financial clout orchestrated through our PAC and key person program, and our much more sophisticated political acumen, began to have an effect. We suddenly became a force to be reckoned with. Soon no legislation that affected accountants was drafted without our input.

LEGISLATIVE AGENDA: RICO AND THE REST

The challenges in Washington during these years were immense. Not only did we have to react to regulatory pressures from Congress and federal agencies like the SEC and FTC, but there were a host of legislative issues we also tried to impact. Our many successes in Washington were a credit to the talented people working in our office there, but also to the thousands of men and women involved in our key person program who took the time to respond when we asked them to call members of Congress and tell them, "Hey, pay attention to our concerns."

Sometimes the Washington office fought for rights we later decided we didn't want. For years, for example, the SEC told us that, notwithstanding a specific exception in the law, CPAs who gave investment advice should have to register as investment advisors. We argued strenuously that CPAs were licensed by the state, so any kind of federal registration was unnecessary. We were concerned about any effort to take away our own self-regulatory prerogatives, no matter what the issue. But Institute members gradually found that registering as investment advisors could actually help their businesses, and we quietly dropped our opposition. In fact, in the mid-1980s Price Waterhouse registered its entire firm as investment advisors. Other firms soon followed suit.

More typical, of course, were intensive legislative efforts to obtain liability relief, one of our top priorities during these years. A prime example were the battles we waged to combat the proliferation of claims against accountants and others under the RICO Act.

RICO was originally intended to allow for recovery of three times damages plus attorney's fees for victims of a pattern of racketeering activity, defined as the commission of two or more criminal acts in a ten-year period. The law listed a number of specific crimes associated with organized crime such as murder for hire, arson, and extortion. But Congress hadn't left it at that. It also added fraud in the sale of securities and wire and mail fraud, which plaintiffs' attorneys eventually broadened to transform virtually any commercial dispute into a RICO claim. In fact, an entire cottage industry emerged of attorneys suing for triple damages under RICO.

The way RICO was being utilized became absurd, and certainly didn't reflect Congress's original intent. It was being used to sue accountants for mail fraud if they mailed a client a financial statement that later was found to include errors. By the end of 1985, we were estimating that lawsuits had been filed under RICO asking for a total of $2 billion against accounting firms, and we expected that number to grow significantly. Making matters worse was that a firm's liability insurance did not typically cover damages won under RICO.

We went through endless rounds of legislative efforts to amend RICO, concentrating most of our resources on attempts to require a prior conviction before a civil RICO action could be taken. That would have meant that an accounting firm or individual CPA could not be sued for multiple damages unless there was a criminal conviction. We helped draft legislation, testified before various House and Senate committees, and at various times came very close to seeing such legislation enacted. In fact, during the 99th Congress remedial legislation failed by just one vote. The Institute also helped organize the Coalition to End Abusive Securities Suits (CEASS), a group of more than 100 companies and trade associations that included the Securities Industry Association, American Bankers Association, National Association of Manufacturers, Chambers of Commerce, the AFL-CIO, and an alliance of insurance companies.

We spent considerable resources arguing for relief before the courts, hiring two influential Washington law firms, Hughes Hubbard and Davis Polk Wardwell. We had some early successes, as lower court judges began to rule that RICO cases must have something to do with organized crime, and that there had to be a prior criminal conviction on a racketeering offense such as murder or extortion before a private party could use RICO to win damages. But in 1985 we suffered a disheartening defeat, when in a decision known as *Sedima v. Imrex*, the Supreme Court ruled in a split 5–4 decision that "although RICO is evolving into something quite different from the original conception, only Congress could correct the problem."

The Institute also began a systematic program of filing amicus (friend of the court) briefs aimed at clarifying the law in cases involving accounting firms and accounting issues. In most of these, some of which went to the U.S. Supreme Court, we were on the winning side. Most importantly, the program helped to develop case law in a way that was fair to the accounting profession and to the public.

The tide began to turn in May 1989, when Supreme Court Chief Justice William Rehnquist published a remarkable opinion piece in the *Wall Street Journal* urging RICO reform. As far as we could tell, never before had a Chief Justice of the U.S. Supreme Court spoken out in such a public manner on pending legislation. Rehnquist wrote, in part:

> The legislative history of the RICO Act strongly suggests that Congress never intended that civil RICO should be used, as it is today, in ordinary commercial disputes far divorced from the influences of organized crime.
>
> I take no position as to which of the reform proposals are acceptable or which is best, but I do think that the imposition of some limitations on civil RICO actions is required.
>
> I think that the time has arrived for Congress to enact amendments to civil RICO to limit its scope to the sort of wrongs that are connected to organized crime, or have some other reason for being in federal court.

Although our efforts to reform RICO were unsuccessful in the legislative arena, clearly we had influenced the debate, and ultimately we did emerge victorious, albeit in the courts. Indeed, in March 1993, after all our legislative attempts had failed, the Supreme Court finally took it on itself to take action. In a 7–2 vote, the high court ruled that accountants, lawyers, and other outside advisors may not be sued under RICO unless they had actually participated in the operation or management of the company.

PRIVATE SECURITIES LITIGATION REFORM ACT

Another intensive legislative effort during these years was aimed at amending the securities acts in order to curb the flood of frivolous lawsuits directed against accounting firms in the 1980s and 1990s. We tried to achieve tort reform at the state level, although this proved to be a slow process. More than half the states did, however, redefine their joint and several liability laws, and privity statutes were passed in several other states. But our first priority was at the federal level. Our Washington office was aided in this regard by the AICPA's Special Committee on Accountants' Legal Liability, established in 1985.

"We spent a lot of time educating ourselves," recalls the committee's chairman, Ray Groves, who was also one of the driving forces behind the Treadway commission (see Chapter Two). "When we began, the concept of joint and several liability was known by maybe one percent of even those decision makers we were trying to reach. We started a process to make the CPA community more knowledgeable about what it could do if it worked together and organized on a single issue. We were particularly successful in mobilizing the key person program to lobby individual lawmakers."

In 1985, the special committee issued a report recommending a series of specific legislative measures to combat the liability crisis, almost all of which were finally passed into law ten years later as part of the Private Securities Litigation Reform Act. Another group working toward these same goals was the National Accountant's Coalition, which the then Big Six firms established to investigate for themselves the liability crisis. "The Big Six decided that if we allowed nature to take its course we'd be sued out of existence," recalls Lee Blazey, the coalition's executive director and a former partner at Arthur Andersen.

Passage of the Private Securities Litigation Reform Act took more than a decade, and was the culmination of the efforts of hundreds of people within the profession. It became law just before Christmas 1995, the only piece of legislation to survive a veto by President Clinton.

The act legislated an auditor's responsibility as recommended by Treadway and others, namely, by requiring the auditor to have in place procedures designed to detect illegal acts and to evaluate the ability of the audited company to continue as a going concern. Auditors were also now required to inform the audited company's management or board of directors if they uncovered illegal acts. If that did not bring an appropriate response, auditors were obliged to report their findings to the SEC. This brought the law in line with the changes we had made in our own rules and procedures. Auditors had always had an obligation to report wrongdoing, but there had been at least two impediments to doing their duty: concern about the loss of a client, and a possible lawsuit for libel or slander. It was the latter concern that was particularly daunting. The change in the law gave a legislative underpinning to the idea that an auditor should disclose wrongdoing, which greatly diminished the threat of a libel suit.

But probably most important from our point of view, the new law replaced "joint and several" liability with a proportionate liability standard. We worked hard to enact this provision so that auditors were now only liable for their fair share of a judgment, corresponding to the percentage of the auditor's responsibility as determined by the court. In other words, if a court determined that an accounting firm contributed 10 percent to the damages incurred by third parties, then the accountants could only be assessed 10 percent of the total monetary damages awarded.

Although the Supreme Court had already provided us with relief from RICO, the reform act also amended that law so that it could only be initiated against those who had been convicted of criminal violations. This had been a central focus of our legislative efforts at RICO reform. An accountant pursued for civil violations could now no longer be accused of racketeering under RICO. The act also curbed frivolous lawsuits, particularly class action suits, by eliminating incentives for plaintiffs to file meritless lawsuits against deep pocket defendants. Primarily this was accomplished by declaring that allegations against accountants had to be specifically spelled out and could only be brought by those who had suffered damages. The *scienter*, or knowledge, standard was clarified so auditors could be held liable under federal security law only if it were determined that the CPA had actual knowledge of wrongdoing. This, we believed, was necessary because accountants had been routinely charged with liability for "aiding and abetting" another person in the commission of fraud by disregarding facts that could have led to its discovery. The reform act also made it easier for courts to prevent frivolous suits because it gave them the authority to charge all court costs to an unsuccessful plaintiff. It also declared that no attorney could serve as lead plaintiff in more than five class action lawsuits within any three-year period.

And finally, the new law gave accountants "safe harbor" for the accuracy of future-oriented information included in financial statements. We considered this important, since without such protection no accountant would ever feel safe making any predictions about the future regardless of how valuable such a prognostication might be to investors.

"None of this was realized overnight, but it just shows what you can do if you're willing to stay the course," recalled Ray Groves.

EDUCATING JUDGES

In addition to legislative reform, the profession was also taking other practical steps to help mitigate the liability crisis. One effort we believed was unique was the course that Don Schneeman, the Institute's general counsel, and his assistant Paul Geoghan, developed aimed at educating members of the judiciary about the responsibilities of CPAs under the securities acts and the rather complicated way accounting and auditing standards were promulgated.

"Marvin Stone, a former AICPA chairman, came up with the idea," recalls Schneeman. "Our concern was that judges and juries who didn't know the first thing about the accounting profession's accounting and auditing responsibilities were being asked to make rulings that profoundly affected the profession. The courses were extremely well received and were even cited in a few court decisions. Some of the judges were astonished, and told us it was the first time someone had offered to teach them something rather than to ask them for something. We knew judges couldn't be lobbied, so we kept it strictly informational, and they appreciated that."

FINANCIAL FRAUD DETECTION AND DISCLOSURE ACT

One of the important lessons we learned as our Washington office became an increasingly effective political force was that lobbying works in two directions. We spent a lot of effort and resources encouraging good legislation, but sometimes it was just as important to forestall what we considered bad legislation. This was particularly true with frequent proposals aimed at taking away our self-regulatory authority and placing standard setting into the hands of the federal government.

One of our biggest challenges was to defeat the Financial Fraud Detection and Disclosure Act, introduced on numerous occasions by Congressman Ron Wyden. On several occasions, particularly during the 99th Congress, this legislation had a real chance of passage. It would have amended the federal securities laws in an effort to ensure that fraudulent activity at major companies would be discovered and reported to the proper enforcement authorities. We lobbied hard against this bill and others like it because we believed they represented a simplistic approach to a complex problem. They would have multiplied the difficulties and expense of auditing, while achieving little public value. They didn't grapple with the truly difficult issue of how the techniques for detecting material fraud could be improved at a reasonable cost.

Moreover, the whistle-blowing aspect of the Wyden bill would have had the effect of transforming the public's watchdog into the government's bloodhound. Not only would it have created an adversarial relationship between CPAs and the entities they audit, it would have required auditors to reach legal conclusions on matters normally outside their area of expertise.

But we also knew that Congress was legitimately concerned that a governing authority be notified when an auditor uncovers fraud. Our old rule was that if a CPA found fraud while conducting an audit, he or she was obligated to tell management and urge them to report it. The rules said nothing about what an auditor should do if management failed to act, but, as already discussed, the rule was later changed so that if management failed to report the fraud, that obligation fell to the auditor. We adopted essentially Congressman Wyden's proposal, but we did it on our own terms, retaining our right of self-regulation.

As we shall see in the following chapter, this became a common theme in the 1980s. On a host of issues, from peer review to CPE, from disclosure of fraud to our rules concerning contingency fees and commissions, the accounting profession recognized that it had to strengthen its rules from within if it wanted to avoid regulation from without.

TAX DIVISION FLEXES ITS MUSCLE

As our Washington office grew more sophisticated, our tax division, under the leadership of Don Skadden, our first vice president of taxation, proved particularly effective. We believed the Institute represented an important voice that should to be heard on any tax proposal being considered by Congress. After all, the Internal Revenue Service is the most pervasive regulatory agency of the U.S. government, with the greatest impact on not only the typical citizen, but also the typical accountant. Today about 25,000 AICPA members belong to the federal tax division, created in 1983. But many more than that are engaged in tax work. As George Anderson liked to say, "scratch a local practitioner and you've scratched a tax man, no matter what they tell you they do."

It might come as a surprise even to most tax practitioners that CPAs working on tax returns is a relatively new phenomenon. Beginning when the national income tax was first implemented by Congress in 1913, and frequently thereafter, the American Bar Association fought to prohibit accountants from becoming involved in tax preparation or planning work.

They argued that it would be paramount to allowing nonlawyers to practice law without a license.

In the end, of course, the accounting profession prevailed, and today, under the leadership of its tax executive committee, the AICPA has about a dozen different tax committees aggressively involved in all aspects of federal and state tax legislation and regulation. Indeed, the AICPA's tax division is kept busy commenting on every proposed change in the tax code—whether it's a minor administrative change promulgated by the IRS, or comprehensive tax reform legislation.

After B.Z.'s arrival in Washington, we made a conscious effort to more aggressively present our views on tax matters. Rather than simply announce whether the Institute favored or opposed a particular piece of legislation, we began to comment on which parts we believed were beneficial and which were ill conceived and why. The modus operandi of our tax executive committee began to change until it was carefully looking at every piece of tax legislation so that it could make specific recommendations to Congress. We became much more sophisticated in joining the political process, offering various alternatives, and becoming a resource for legislators. In fact, by the end of the 1980s the Institute's tax division had become so highly respected for its expertise and objectivity that it was often requested by Congress to comment and testify about pending legislation. This new strategy also increased our clout with the IRS, an agency with which many of our members interacted on a regular basis.

Prior to the mid-1980s, we also had been reluctant to lobby Congress concerning any issue in which we had an obvious self-interest. This too changed. One reason we became so actively involved in the discourse surrounding the Tax Reform Act of 1986, for example, was that its year-end conformity clause would have compressed audit and tax work into an even smaller period of time. This wasn't so much an issue for our clients as it was for practitioners, and we fought successfully to remove it from the bill.

By the May 1990 Council meeting, Arthur Hoffman, chairman of the AICPA's Federal Taxation Executive Committee, was able to inform Council members that the sophistication of our tax division had reached the point where it provided Congress better and more objective positions than virtually any other organization. "When the roundtable approach to developing legislation is used, I believe it is now recognized by Congress, the Treasury, and the IRS that we deserve a seat," Hoffman said. "Regardless of the subject matter, we can put in that seat the most knowledgeable tax experts in the country, with unmatched staff support."

Often our tax division provided congressional committees with its expertise without taking a position on a particular issue. During the various debates over the years concerning the flat tax, for example, the AICPA did not take a position either for or against any specific proposal. But we did thoroughly investigate the impact a flat tax would have on various segments of the taxpaying public, then shared that analysis with the appropriate congressional committees.

One theme that the tax executive committee sounded continuously over the years was tax simplification. While we generally did not draft specific legislation on our own, the Institute's position has always been very clear. The current complexity of the tax code has become increasingly grotesque and needs to be reformed. In fact, the tax code, which runs an incredible 10,000 pages, is the most complicated of any nation in the world. That's a status quo that simply cannot be defended.

On the surface, it might seem that the accounting profession would applaud an intricate tax code, since ostensibly the more complicated the process of tax preparation, the more work there will be for tax preparers. (Cynics have been known to call tax "reform" legislation "the accountants and lawyers relief act.") But nothing could actually be further from the truth. For as long as I can remember—and certainly throughout my tenure at the AICPA—the Institute, led by a committed and knowledgeable tax committee, took our public responsibility seriously when it came to commenting and advising on tax matters.

"At the end of the day, the tax committee tries to take the most publicly responsible position," agrees Dominic Tarantino, a former chairman of the AICPA board and a longtime observer of the federal tax scene. "We try to be constructive, while at the same time pointing out the infirmities of any proposal."

Another of the Institute's priorities during these years was passage of federal legislation that would put the confidentiality of tax information disclosed to a CPA on par with information provided to one's attorney. We maintained that the attorney–client privilege gave lawyers a professional advantage over CPAs, since clients might be reluctant to discuss sensitive tax matters unless protected from disclosure to the courts.

CPAs were finally given this confidentiality protection in July 1995, and then only because well-publicized abuses by the IRS against individual taxpayers prompted Congress to pass taxpayer rights legislation. Even then we didn't get everything we had hoped. Our confidentiality privilege does not apply, for example, to tax shelters or criminal matters, so it still is not equal to that given attorneys. (One way a taxpayer can obtain full confidentiality on all issues is to hire an attorney, who can in turn hire an accountant. Under a principal called "attorney work product," this in effect "cloaks" the tax expert with the broader attorney–client privilege.)

SELF-REGULATE BEFORE WE REGULATE YOU OURSELVES

It was all well and good to make our mark in Washington by impacting legislation and warding off congressional efforts to regulate our profession. We could point to plenty of successes during these years that we could be proud of, not the least of which was our construction of a Washington office into an effective political force.

But we also recognized that much more than political action would be needed if we were to prevent Congress or the SEC from taking measures that would remove our own self-regulatory control and place it in the hands of the federal government. We knew we would have to strengthen dramatically our internal rules and controls. The following chapter will discuss the extraordinary measures the AICPA took to make certain that the profession, and no one else, remained in control of its own destiny.

Chapter 4

Regulate Thyself:
Standards of Professional Conduct

As we have seen, by the mid-1980s, spurred by high-profile bank failures and corporate bankruptcies, Congress, the SEC, and the FTC were all putting enormous pressures on the accounting profession to accept additional responsibilities or be prepared to have the federal government take those responsibilities away from us. Within this environment, and in recognition that change was necessary, in October 1983 the Institute appointed a Special Committee on Standards of Professional Conduct to recommend a course of action. Once again we relied on the Future Issues Committee to suggest specific topics that needed to be addressed, namely:

1. Expansion of services and products
2. Changes in the nature and extent of competition in the profession
3. The role of self-regulation
4. Improving the quality of practice
5. Independence and objectivity

SPECIAL COMMITTEE ON STANDARDS OF PROFESSIONAL CONDUCT

Rholan E. Larson, then board chairman of the AICPA, and I were fortunate to be able to persuade George Anderson to serve as chairman of this new special committee, which is generally recognized as one of the seminal events in the history of our profession. George, then managing partner of Anderson ZurMuehlen & Co. of Helena, Montana, and a former AICPA chairman (1981–82), was the ideal person for the job. We already had worked together on establishing the strategic planning process (see Chapter One). Rholan and I knew he could be entrusted with the special committee's vital charge of "restructuring professional standards to achieve professional excellence in a changing environment" in a way that would satisfy both public demands and professional needs. Indeed, in the opinion of many, the report of the special committee that was issued three years later represented the most thorough reassessment of ethical guidelines that the profession had ever made, and certainly the most meaningful since the 1973 restatement of the Code of Professional Ethics.

George understood the necessity for a real transformation, not just cosmetic change. In a report to the October 1985 Council meeting, he told us, "The committee has evaluated the present code and found it wanting. Commercialism is winning out over professionalism. It is our opinion that the present code does not encourage professionalism like it should, and that it is very difficult to enforce."

And, as he wrote in the *Journal of Accountancy* not long after the committee's final report was issued, "Although much of the more strident criticism is unwarranted, clearly we cannot emulate the ostrich or duck our heads to avoid the scattershot. Such behavior ill suits a serious profession— nor in the end would it work. The public's legitimate expectations must be reckoned with; the profession's response must be visible and, to be credible, meaningful."

In the end, Anderson believed that his committee's report and recommendations could "breathe new life into concepts that the profession has always valued: the public interest, integrity, independence and objectivity, due care—in sum, the highest attainable level of professional service and conduct."

The report of the special committee (dubbed the Anderson report) was presented at the 1986 spring AICPA Council meeting. Essentially it called for a three-pronged attack on the possibility of substandard work: better education of accountants at the entry level, as well as continuing, lifelong education; peer review of accounting and auditing practices to help ensure that these services were being provided according to the high principles established by the profession; and a revised code of professional conduct supported by a more effective enforcement process.

During the three years that the Anderson committee took to prepare its report there was some concern, and impatience, about the length of time it was taking to complete the study. But I think most agreed that the result was worth waiting for. It combined practical steps for improving performance, most notably better-trained people, with more effective ethical standards and better enforcement of rules to prevent substandard work. Many of us at the Institute were convinced that the committee recommendations, if adopted, would result in some of the most important changes in our profession's history.

STRAIT COMMITTEE

Our next step, therefore, was to make certain that the Anderson committee recommendations did not just sit idle. Toward that end, an implementation committee was appointed, chaired by AICPA Vice Chairman Marvin Strait, managing partner of Strait, Kushinksy & Co. of Colorado Springs, Colorado. Its charge was to consider members' views on the proposed changes and to develop a detailed plan for their adoption. Specific proposals would then be submitted to a membership vote.

The Strait committee began by making a grass-roots effort to elicit support from our membership. With the help of every state society, it conducted focus groups in each state to ask members what they thought of the Anderson committee recommendations. Marvin and the other committee members worked hard to set up a partnership with the state societies. Their efforts resembled a national political campaign, as we all knew how important it was to get support at the most local level. We had to demonstrate to our national membership that the Anderson committee recommendations would be beneficial not only to the collective profession, but also to each individual member's livelihood.

"I think it helped that I had a small-town perspective," Strait recalls. "I gave speeches all over the country in favor of the proposals. Oftentimes a local practitioner would come up to me and say, 'You guys from New York don't understand the real problems of accountants outside the big cities.' I think it gave me more credibility when I could explain I was one of them, that when I started practicing in the 1960s in rural Colorado, about 200 miles southeast of Denver, I was the only CPA within five counties. You can't get more rural than that."

More than 7000 AICPA members participated in the implementation committee's focus groups. Hundreds more members expressed their views in other ways, mostly through letters to the committee or to the AICPA's governing Council. The implementation committee then used these opinions to refine and modify the Anderson report to form a *Plan to Restructure*

Professional Standards, the elements of which we successfully submitted to a membership vote.

Most of the Anderson committee recommendations, with the notable exceptions being the sections on contingent fees and commissions, were included in the final plan. In the spring of 1987, the AICPA governing Council overwhelmingly endorsed it and strongly urged our membership to vote yes on each of six specific proposals:

- Proposal One: To adopt a new code of professional conduct
- Proposal Two: To adopt as a membership requirement quality review for members in public practice
- Proposal Three: To restructure the Joint Trial Board to streamline the AICPA's enforcement procedures (This was implemented in order to establish more effective procedures for handling complaints by redefining the roles and responsibilities of the AICPA and the state CPA societies in the Joint Ethics Enforcement Program.)
- Proposals Four and Five: To adopt as membership requirements CPE for both those in public practice and not in public practice
- Proposal Six: To adopt as a requirement that, after the year 2000, applicants for AICPA membership must have at least 150 collegiate-level semester hours, including a bachelor's degree or its equivalent

Many people both inside and outside the profession told us we didn't have a chance of passing the *Plan to Restructure*'s most controversial recommendations. Particularly dicey were the two new requirements for mandatory peer review and CPE. At more than one Council meeting, members expressed the concern that if the Anderson report were implemented it could cause a substantial erosion of AICPA membership. Figures of 10 percent; some as high as 25 percent were routinely bandied about. Many Council members were particularly fearful that industry members would object to the cost of CPE and wouldn't recognize how it could benefit them.

But much to the Council members credit, the prevailing view was that we would have to accept any negative consequences that might occur. Regardless of the repercussions, it was vitally important that these new membership requirements raise standards and give increased stature to AICPA membership. We communicated this message to members in a marketing and communications program directed specifically to them. Taken together, the *Plan to Restructure* represented a radical change for the accounting profession. The code of professional ethics was revamped, peer review was expanded and made mandatory, and, as we shall see in the following chapter, the educational requirements for all AICPA members were significantly strengthened.

CODE OF PROFESSIONAL CONDUCT

The first proposal of the *Plan to Restructure*, a new code of professional conduct to replace the existing code of professional ethics, may not have been the Anderson committee's most controversial recommendation, but that did not negate its significance. One of the unique characteristics of a profession is the existence of a code of conduct to guide behavior and measure performance. As far back as 1905, the American Association of Public Accountants (the first professional accounting association and a precursor of the AICPA) adopted an initial set of ethical rules. In 1917 another AICPA predecessor, the renamed Institute of Accountants, established a code of ethics with eight rules. Some were rather nebulous, such as a warning against "engaging in activities incompatible with the practice of public accounting." Others were more specific, including a prohibition against "soliciting clients of other members of the Institute."

As the profession's code of conduct evolved, adjusting as it must to the ever-shifting business environment, two principles remained paramount. First, the code was designed to provide guidance to CPAs so they could maintain high standards of integrity and objectivity. And second, ethical standards were made flexible enough to allow accountants the opportunity to earn a livelihood. These two forces—the need for accountants to maintain integrity and objectivity in the face of an increasing number of specialized commercial business opportunities that opened up to the modern-day accountant—have always formed the essence of the tug of war the profession's code of ethics has tried to mediate.

The restructured code, as recommended by the Anderson committee and ratified by the AICPA membership, consisted of two sections: principles and rules. The principles were aspirational and goal-oriented, but not enforceable. The rules established minimum levels of behavior and provided a basis for disciplinary action.

The first section contained "positively stated principles" that provided the framework for the profession's rules "that prescribe the ethical responsibilities members should strive to achieve." These new standards, the Anderson report declared, "require an unswerving commitment to honorable behavior even at the sacrifice of personal advantage."

The standards were intended to be applicable to all members of the AICPA, whether in public practice, industry, government, or academia, and to "all their professional responsibilities." They reaffirmed the "essential role in society" played by CPAs and further clarified a CPA's responsibilities. According to the Anderson report, "Members have a continuing responsibility to cooperate with each other to improve the art of accounting, maintain the public's confidence, and carry out the profession's special responsibilities of self-governance."

It went on to state that "in discharging their professional responsibilities, members may encounter conflicting pressures from clients, employers, and the public at large. In resolving those conflicts, members should act with integrity, guided by the precept that when members observe their responsibility to the public, clients' and employers' interests are best served."

Emphasizing the CPAs' responsibility to the public, the new code explained that the profession's "public" includes "clients, credit grantors, governments, employers, investors, the business and financial community, and others who rely on the objectivity and integrity of certified public accountants."

The importance of integrity among members was also reaffirmed. "Integrity is measured in terms of what is right and just," the report declared. "Service and public trust should not be subordinated to personal gain and advantage .... a member should test decisions and deeds by asking: 'Am I doing what a person of integrity would do? Have I retained my integrity?'"

The new code acknowledged the distinction between members who provide auditing and attest services and those involved in other management or consulting capacities. Still, integrity and objectivity were critical when performing all professional services. "Such a member who provides auditing and other attestation services should be independent in fact and appearance. In providing all other services, a member should maintain objectivity and avoid conflicts of interest."

Acknowledging that although members who were not in public practice, (corporate in-house CPAs, for example) "cannot maintain the appearance of independence...they nevertheless have the responsibility to maintain objectivity in rendering professional services."

And finally, in an effort to address criticism of the substandard work of a few engagements, the new code made it clear that a high level of competency was demanded of all AICPA members. A section entitled "Due Care" stated, "In all engagements and in all responsibilities, each member should undertake to achieve a level of competence that will assure that the quality of his or her services meets the required high level of professionalism."

SCOPE AND NATURE OF SERVICES

One of the more contentious areas the Anderson committee had to address was the scope and nature of nonattest services a CPA should offer. Since the work of professional CPAs now reached far beyond the core business of auditing and tax preparation, the Anderson committee needed to establish some general guidelines that members could consult in their pursuit of other work, both from their auditing clients and from others. Since the principles section of the Anderson committee report was meant to provide AICPA members overall guidelines of conduct, the report acknowledged that "no hard-and-fast rules can be developed to help members reach these judgments. A member must be satisfied that standards of professional conduct are being adhered to in this regard." It was up to the individual member to determine whether in his or her judgment, "the nature and magnitude of other services provided to an audit client over time might create or appear to create conflicts of interest in the performance of the audit function for that client."

The report recognized that maintaining independence and freedom of conflicts of interest may "in some instances...represent an overall constraint on ... non-audit services that might be offered to a specific client over time." Members were required to "justify any departure" from these standards. Compliance would be reinforced "ultimately by disciplinary proceedings, when necessary, against those who fail to comply with the profession's adopted performance standards."

RULES OF PROFESSIONAL CONDUCT

In the rules of performance and behavior section of the proposed new code of ethics as ratified by our membership, the Anderson report reaffirmed the AICPA's commitment to independence (Rule 101), integrity and objectivity (Rule 102), general standards of professional competence (Rule 201), compliance with standards (Rule 202), and conformity with generally accepted accounting principles (Rule 203). Rule 301 reaffirmed the commitment to the confidentiality of client information, while still requiring members to comply with legal requirements.

Independence (Rule 101), advertising, competitive bidding and solicitation (Rule 502), and contingent fees and commissions (Rules 302 and 503) form the three areas of professional conduct that are at the heart of the Institute's code. They became topics of great debate and interest to the profession and the public, and each of them warrants a more detailed discussion.

INDEPENDENCE

No concept is more fundamental to the idea of a CPA earning the trust of its various constituencies than independence. After the 1929 stock market crash, the independence of auditors was increasingly called into question, as it would be periodically during the ensuing decades. As a result, the AICPA embarked on a reevaluation of standards, leading to a 1931 resolution stating that "the maintenance of a dual relationship, as director or officer of a corporation, while acting as auditor of that corporation, is against the best interests of the public...and tends to destroy the independence...considered essential in the relationship between client and auditor."

In 1933, the Institute adopted a resolution prohibiting its members from holding a financial interest in a publicly financed enterprise for which he or she served as an independent auditor. But eight years later, a newly adopted rule provided that a member could have a substantial financial interest in a client that was privately owned and still be independent if the extent of the member's financial interest was made public. But then in 1964, the Institute membership accepted a new standard that any direct or material indirect financial interest or service as a director, officer, promoter, or key employee of a public *or* private client would be considered compromising to a member's independence toward that client.

The code of professional ethics adopted in 1973 reaffirmed this view, as did the Anderson report, which stated that "a member should maintain objectivity and be free of conflicts of interest in discharging professional responsibilities. A member should be independent in fact and appearance when providing auditing and other attestation services." The report stressed that independence and intellectual honesty were crucial in the performance of attesting and auditing functions.

As did previous versions of the code, the new rules as recommended by the Anderson committee spelled out examples of independence that would be considered to be impaired. One criterion, for example, was whether during the period of professional engagement the auditing firm "had or was committed to acquire any direct or material indirect financial interest in the enterprise."

Increasingly in the 1980s, as the largest accounting firms began to concentrate on nonattest consulting services, independence of these firms was questioned by Congress and other regulatory authorities. When I represented the Institute before Congressman Dingell's subcommittee on oversight and investigations in March 1985 (see Chapter Three), my testimony tried to address head on the subcommittee's skepticism as to whether an accounting firm that acts as an advocate in helping a client with management or tax decisions can simultaneously fulfill its public responsibilities by exercising independence in auditing the firm's financial statements. Their concern was that auditors might look the other way if they had a strong financial interest in pleasing the client. This was a crucial issue, since several subcommittee members had already endorsed legislation that would have banned accounting firms from providing both management consulting and audit services for the same client.

At the same time, accounting firms were just beginning to become multifaceted global enterprises, something that today is recognized as essential to our global financial, information, and investment infrastructure. A prohibition of providing consulting services to attest clients would have had a disastrous effect on many firms. It also would have made it extremely difficult to recruit the best and brightest talent available.

But the idea that large consulting fees somehow threaten the independence of the audit has never made much sense to me. First, after extensive study by groups both inside and outside the profession, no empirical evidence had ever been found to demonstrate that the provision of MAS undercut the independence, quality, or objectivity of audits. In fact, providing management consulting services often helps auditors understand their client's business better, and therefore improves their ability to conduct a thorough audit. Besides, the threat of litigation, combined with the loss of the firm's reputation, was more than enough of a safeguard. (Ironically, during the congressional hearings we actually played down the litigation threat as a deterrent because we were simultaneously pushing for legislative relief from frivolous lawsuits.)

Moreover, the size of the fee for consulting services was really a bogus issue. Even if a large consulting fee was going to tempt an auditor to knowingly issue a false audit report or purposely hide a company's financial difficulties, the temptation would be equally strong because of the audit fee alone, which for large corporations can run into many millions of dollars.

In my former life as a partner with Main Hurdman responsible for accounting and auditing issues, I was more concerned with the ability of audit partners to stand up to tough clients than I was with large consulting fees affecting their judgment. If anything is going to make an auditor bend the rules, it is the intimidation factor—the inability of an individual auditor to withstand pressure from an aggressive client. As my former managing partner, Leroy Layton, liked to say, "A firm is like a link chain, only as strong as its weakest partner." Lee later served as chairman of the Institute's Accounting Principles Board and chairman of the board of the Institute.

OTHER STUDIES ON INDEPENDENCE

In addition to the Anderson committee and the Kirk report (see Chapter Two), the accounting profession conducted a number of other studies on the issue of auditor independence during these years. In 1986, for example, two different surveys were conducted to determine public attitudes. First, the POB surveyed key members of the public, which included groups of businesspeople and financial analysts. The purpose of the study was to determine the effect, if any, MAS had on the perception of auditor independence and objectivity.

The POB survey found that many of the nonaudit services performed by CPAs—everything from designing computer systems to providing actuarial services for a company's pension plan—were not considered to compromise an accountant's independence and objectivity. Certain other services, however, were considered harmful. These included identifying merger or acquisition candidates, employment searches for senior management positions, setting values on acquired assets during business mergers, and designing executive compensation plans.

Another survey, this one conducted by the AICPA, was designed to measure the general public's attitudes toward the profession. Its findings confirmed those of the POB survey in that they indicated that many nonaudit functions by CPAs were considered appropriate to the profession, while other services, such as appraisal, executive search, and selling tax shelters, were not.

OPINION SHOPPING

The AICPA has always tried to protect the status of accountancy as a profession, defined by *Webster's Dictionary* as "a calling requiring specialized knowledge and often long and intense academic preparation." One way the profession attempted not only to maintain its independence from clients but also to distinguish it from ordinary business- or salespeople, was to prohibit certain sales techniques. These include opinion shopping, commissions, contingent fees, competitive bidding, and even advertising.

Opinion shopping, for example, refers to a practice whereby a client who has a disagreement with its auditor shops around to find another auditor who will agree with the client's position. The obvious threat is that the current auditor could be fired, which places enormous pressure on the firm to alter its position.

By the mid-1980s, an increase in competition within the profession, combined with the proliferation of highly complex financial transactions, increased the temptation on the part of many companies to opinion shop. Indeed, according to data accumulated by the SEC, the number of public companies that changed auditors almost doubled between 1984 and 1986. And according to the accounting newsletter, *Public Accounting Reports*, between 1981 and 1985 about 20 percent of auditor changes occurred after a client had received a qualified report from its auditors.

Within this kind of environment, it was no surprise that the federal government began cracking down against the practice. In fact, a major reason that the Justice Department sued Peat Marwick Mitchell & Co. in July 1985 for giving Penn Square Bank an unqualified report only months before the bank went bankrupt was that it believed the firm had been influenced by the fact that a competitor, Arthur Young, was fired after issuing a qualified report in 1980. The SEC also accused Broadview Financial Corp. of Cleveland of firing Peat Marwick when it wouldn't interpret a real estate transaction in a way that would improve the company's bottom line. The SEC acknowledged that there was nothing necessarily improper about changing auditors, but "auditors should act cautiously when approached by an issuer in search of an opinion which differs from that given by its existing auditors."

In another example, in 1986 executives at Silverado Banking Savings and Loan of Denver, Colorado, hired a new auditor after reporting losses of $20 million in 1985, mostly due to problem loans. Silverado subsequently reported a dramatic turnaround in 1986, so much so that its $15 million in profits allowed top executives to take home $2.7 million in bonuses. But eighteen months later the company collapsed, at a cost to the federal government of an estimated $1 billion.

The AICPA was caught in the middle during these years. On the one hand, we recognized that opinion shopping was a potential problem that needed constant monitoring. On the other hand, just as second medical opinions could effectively be used to improve quality of care and to control costs, second accounting opinions, if not abused, could be a legitimate tool for inquiring into the accounting of a complex transaction. This was particularly true in an era in which GAAP was constantly evolving and increasingly complex.

As a result, the AICPA leadership believed that outlawing opinion shopping altogether was not the solution. We certainly didn't object, however, when in 1988 the SEC expanded its requirements that companies inform the SEC whenever it changed auditors. Companies were now required to indicate if there had been a disagreement with the former auditors. If there had been, they had to indicate whether the auditor had resigned or been fired, whether the disagreement had been discussed with the company's board of directors, and whether management put limitations on communications between the new and old auditors.

We were also pleased, however, that the SEC backed down from its initial proposal that would have forced disclosure of every discussion companies had with their auditors. That, we believed, would have had a chilling effect on the confidence companies had in discussing their financial matters openly and honestly with their accounting firms.

ADVERTISING, COMPETITIVE BIDDING, AND SOLICITATION

Another historically contentious ethical area covered by the Anderson committee was advertising. Until 1978, a general ban on both advertising and direct or indirect solicitation of clients continued to be in effect. The 1973 restated ethics code stated simply, "A member shall not seek to obtain clients by solicitation. Advertising is a form of solicitation and is prohibited."

This wasn't unusual for professional groups. The legal profession, for example, didn't allow advertising either, until the U.S. Department of Justice in the 1970s objected to what it considered anticompetitive provisions in the ethics codes of a number of professional organizations. The congressional hearings of the late 1970s chaired by Senator Lee Metcalf (see Chapter Three) specifically criticized the accounting profession's efforts to ban advertising. And then a 1977 landmark Supreme Court decision settled the issue. In *Bates v. State Bar of Arizona*, the Supreme Court ruled that "professional organizations were not exempt from the federal antitrust laws and that such organizations' prohibitions on advertising were contrary to antitrust laws."

As a result of these and other events, in 1979 the AICPA lifted its ban. It did, however, continue to restrict certain forms of advertising, such as self-laudatory and comparative claims, testimonials and endorsements, and advertising not considered professionally dignified. Advertising and other forms of solicitation that were "false, misleading, or deceptive" were still prohibited.

Despite the lifting of a total advertising ban, the continued prohibition against direct, uninvited solicitation of a specific potential client no doubt discouraged many accounting firms from launching full-fledged advertising campaigns, even if they were inclined to do so. Largely as a result of pressure from the FTC, in 1983 the Institute's ban on direct, uninvited solicitation was also removed and replaced with "a member shall not seek to obtain clients by advertising or other forms of solicitation in a manner that is false, misleading, or deceptive. Solicitation by the use of coercion, overreaching, or harassing conduct is prohibited." In essence, once the profession was pushed into allowing advertising, we tried to ensure the truthfulness of such advertising.

As a result, by the time the Anderson committee had issued its final report, the issue of whether to allow accountants to advertise had been largely resolved. Our new Rule 502 as contained in the *Plan to Restructure* was only slightly different from the previous language.

In Ethics Interpretation 502-2, the AICPA provided examples of specific advertising that was still prohibited. Most of them were taken directly from the Anderson report and included advertisements that:

- Create false or unjustified expectations of favorable results
- Imply the ability to influence any court, tribunal, regulatory agency, or similar body or official
- Consist of self-laudatory statements not based on verifiable facts
- Make comparisons with other CPAs that are not based on verifiable facts
- Contain a representation that specific professional services will be performed for a stated or estimated fee when it was likely that such fees would be substantially increased and the prospective client was not advised of that likelihood
- Contain any other representation that would be likely to cause a reasonable person to misunderstand or be deceived

But even these rules were soon under attack by the federal government on antitrust grounds, and as a result, by 1992 our ethics code allowed CPAs to use any form of advertising. Testimonials, endorsements, self-laudatory advertisements, direct comparisons of one firm to another, and in-person solicitation of prospective clients were all permitted. Even undignified advertisements or those in poor taste were allowed, although false, misleading or deceptive statements were still prohibited.

None of these relaxations of our rules, including our initial 1979 advertising ban, sat particularly well with much of our membership, as represented by the Council. In fact, during the early 1980s the Council searched for ways to reinstate the prohibition. At our October 1980 meeting, for example, we voted to appoint a special committee to study the ramifications and legality of current rules pertaining to direct, uninvited solicitation. As part of its work, that committee commissioned a survey that found that about two-thirds of our membership favored the reinstatement of a ban on direct, uninvited solicitation because they believed it harmed the professional image of CPAs and was not in the public interest.

The committee also began a search for a law firm with antitrust expertise, finally settling on Kay, Scholer, Fierman, Hays and Handler. The question we posed to them was, "Can a broad rule banning solicitation be reinstated in the AICPA Code of Professional Ethics, and if so, what is the strongest rule we could establish?" At the very least, we wanted to know if we could issue a policy statement condemning direct uninvited solicitation, even if it would be prophylactic and unenforceable.

In response, the law firm prepared for us a 122-page opinion telling us we could not go back to the old rule without incurring a substantial antitrust risk. It wasn't the conclusion we had hoped for, of course, but our lawyers made it very plain that should we try to reinstate a general prohibition against direct, uninvited solicitation, it was likely that an adverse court decision would not only strike down the rule, but issue a decree that would probably restrain the AICPA from ever expressing its views on the subject again, and perhaps strike down other rules as well. After all, the courts had already ruled that an advertising ban by both doctors and architects violated the Sherman Antitrust Act. At least under current law we could retain our rule prohibiting members from seeking to obtain clients "by advertising or other forms of solicitation in a manner that is false, misleading, or deceptive" or "solicitation by the use of coercion, overreaching, or harassing conduct."

So, reluctantly, the Council acquiesced to the inevitable, although not happily. Indeed, at our October 1981 meeting, the Council overwhelmingly passed a resolution that "it is the sense of this meeting that the AICPA deplores the practice of direct, uninvited solicitation of clients by its members and regrets that, due to the advice of counsel, an appropriate ban on this practice will not be part of the Code of Professional Conduct."

Two years later the bitterness still remained. "It is my personal view that the continued erosion of what was once thought of as appropriate behavior for professionals creates an atmosphere of commercialism, as opposed to professionalism, which in my view is not in the best interests of the profession or the public," declared NASBA president Robert Block at our May 1983 Council meeting. "The present pell-mell campaign to grab the other fellow's client at almost any cost threatens to change the rendering of professional opinions on financial statements from a professional service to a commodity. I believe that the outside forces which are attempting to further erode our behavioral prohibitions are creating an environment which could seriously damage or destroy the status of professionalism which we have fought so long to attain."

CONTINGENT FEES

Between 1980 and 1995, no issue was more contentious than the battle over contingent fees and commissions. A contingent fee is a fee for service based on the result of the service rendered. An accounting firm, for example, is being remunerated on a contingency basis if it isn't paid unless it comes up with a predetermined result. A commission is simply a fee paid either for referring a client to another professional, or for suggesting that a client use or purchase a product from a third-party vendor—accounting software, for example.

The reason for discouraging these practices was self-evident, at least it was to most members of the accounting profession. If a CPA's fee is determined by the outcome of an engagement, how will the CPA be able to maintain independence from the client's interests? Remember that independence and objectivity have always been the hallmarks of our profession.

As early as 1919, the AICPA adopted a rule prohibiting contingent fees. In 1936 it clarified the policy, noting that certain services, such as tax returns, could be performed on a contingent fee basis. These exceptions were further clarified in the 1973 restatement of the Code of Ethics, but a general prohibition was retained.

Prior to the Anderson committee recommendations being approved by membership, Rule 302 stated that "professional services shall not be offered or rendered under an arrangement whereby no fee will be charged unless a specified finding or result is attained, or where the fee is otherwise contingent upon the finding or results of such services."

The Anderson committee recommended that Rule 302 be changed to state only that "a member in public practice who performs engagements for a contingent fee would be considered to have lost independence with regard to that client because a common financial interest has been established."

This in effect would have limited the rule's application to clients for whom engagements requiring independence were performed. The Anderson committee was trying to walk the tightrope between the critical importance of maintaining auditor independence in traditional auditing and attestation areas, and the growing recognition that contingent fees were appropriate in the selling of certain CPA services.

The most important rationale for the change was as a result of the expansion of services offered by accounting firms. They were now often competing with nonaccountants, who were not restricted by these kinds of obstacles. An overall prohibition of contingent fees was simply unfair to those members involved in engagements for which a contingent fee might be the most appropriate arrangement available—Medicare reimbursement engagements, for example.

In the case of contingency fees, however, Marvin Strait's implementation committee received so many negative comments to the proposed changes to Rule 302 that in the end it was left unchanged. Accountants across the country weren't about to give blanket approval for any fee arrangement a client might suggest as long as it was not for audit work. Their objections went beyond contingent fees. Another Anderson committee recommendation, that "bodies designated by the AICPA governing Council" would be permitted "to determine circumstances in which commissions would be acceptable," also was omitted from the final restructure plan.

COMMISSIONS

Commissions, many believed, impair a member's objectivity and mar the public perception of the CPA as a professional. As a result, the 1973 restatement of our ethics code continued to prohibit "payment of a commission to anyone to obtain a client, and receipt of a commission from anyone for referral to a client of products or services of others."

The new Rule 503, as prepared by the Anderson committee, stated that "the acceptance of a payment for referral of products or services of others by a member in public practice is considered to create a conflict of interest that results in a loss of objectivity and independence except under those circumstances where bodies designated by Council have determined that such conflicts of interest do not arise." In effect, then, the new rule would have lifted the outright ban and put the burden of decision in the hands of committees so designated by the AICPA.

As George Anderson wrote in the *Journal of Accountancy*, "The proposed rule would explicitly give the AICPA senior technical committees a key role to play: Through their interpretations, they could determine circumstances in which accepting a commission wouldn't create a conflict of interest that results in a loss of objectivity and independence."

But as with contingent fees, the Anderson committee's recommended changes to Rule 503 regarding commissions were not included in the Code of Professional Conduct submitted to a membership vote in 1987. The feeling among AICPA members was simply too strong that a complete lifting of these bans was a slippery slope leading inevitably to the loss of professionalism.

Yet many of us understood that the prohibition against contingent fees and commissions could not be sustained. Neither the marketplace nor the federal government would allow it. In the first place, the expansion of the types of services provided by CPAs, in part caused by an effort to supplement income from their dwindling core attestation activities, was making inevitable a broadening of the manner in which CPAs would be compensated for their work. Faced with increasing competition, charging on an hourly basis alone was not going to appeal to many potential clients. In reality, exceptions were already regularly being granted—for client referral between accountants, for example.

But there was another, perhaps even more important consideration as we debated whether to allow contingent fees and commissions. By the mid-1980s, the FTC had begun to object to the restrictions because it believed they lessened competition and inhibited free commerce.

FTC VERSUS AICPA

The AICPA spent considerable resources during these years responding to the various congressional committees and subcommittees that seemed always to be investigating the accounting profession. Nevertheless, it was actually the FTC that put the most pressure on us to change our rules and regulations. In fact, to those who had been active in AICPA affairs for many years it probably seemed as if there was never a time when our ethics rules were not under antitrust attack. In the most publicized instance, in the late 1960s the U.S. Department of Justice issued a legal complaint aimed at overturning our ban on advertising. Two highly qualified antitrust law firms warned us that if we tried to enforce the rule against a member, the member's defense would likely be based on the illegality of the rule and the member could successfully sue to prevent enforcement. As a result, in 1972 the Council authorized the AICPA to accept a settlement so that the only restriction on advertising that remained in our bylaws was that it could not be false, misleading, or deceptive.

That capitulation was still fresh in the minds of many AICPA Council members as the federal government set their sights on our prohibition of commissions and contingency fees. Contingent fees became a major issue in 1984, when at our April meeting, the AICPA's Board of Directors recommended that the Council approve a mail ballot of the membership to modify Rule 302 so that contingent fee arrangements would be prohibited in independence-related engagements only. The existing rule prohibited such fees in *all* professional engagements.

In making this recommendation, the Professional Ethics Division and the Board of Directors relied heavily on the advice of the Institute's outside counsel at Willkie, Farr & Gallagher, who warned us that by keeping the total ban on commissions we were opening ourselves up to litigation by members who felt our prohibition violated antitrust law. Our attorneys also feared it was only a matter of time before the federal government took that same position.

At our October 1986 meeting, board chairman Marvin Strait asked Council members, "Are we better off trying to fashion a rule that can stand a court test and we can keep, or are we in fact better off fighting and letting someone else set it for us?"

The Council, however, was not convinced. Don Nelson, speaking on behalf of the California Society of CPAs, argued that "our profession is placed above others because our services are rendered without bias. Our integrity is based to a great extent on the fact that we do not accept commissions or contingent fees. If the courts were to mandate that we must accept these things, then so be it. But let's let the courts make that decision and not make it voluntarily ourselves. We think it's time to draw the line and stand up for what we think as public servants is proper. We should not be intimidated by less important arguments that, in the final analysis, could destroy the backbone of our profession."

After extensive debate, the Council refused to authorize a mail ballot. As a result, in July 1984 the AICPA's board of directors lifted its suspension of enforcement of Rule 302. It soon decided to revisit the issue, however, after the Colorado CPA society's leadership voted unanimously to request the Institute to again bring before the Council the question of a membership vote on the contingent fee rule.

At the October 1984 Council meeting, Council members expressed their frustration that the issue was once again back on their agenda. "I would argue that most of us in this room, if given the opportunity, would gladly take back the vote we cast a few years ago to remove our prohibition on direct and uninvited solicitation," declared Bob Ford, a Council member from California. "We've seen where that action has taken our profession; the damage it has done both internally and in the eyes of the public. Hopefully we will not repeat that mistake. If we are to continue to gradually chip away at our behavioral standards, then I suggest we simply abolish all standards and tell our members the door is wide open, that anything goes."

As a compromise, other Council members felt that any action on contingent fee rules was premature, since the Special Committee on Standards of Professional Conduct had been working for the past year to review the AICPA's entire ethical code. These members felt that any Council action should await the Anderson committee's report.

So again, the Council could not be persuaded to seek membership approval of a rule change on contingent fees. But then just as the Institute's board of directors had feared, a few months later the other shoe dropped. In January 1985, the FTC's Bureau of Competition notified the AICPA that it had begun a preliminary investigation into certain ethics rules and interpretations of our code of professional conduct. We subsequently received a request for the production of documents.

Immediately the AICPA agreed to voluntarily cooperate with the investigation. It wasn't as if we had a lot of choice in the matter. The alternative would have been to be forced to produce the documents by subpoena. Besides, this way we could at least try to furnish these documents on our own terms.

Producing the requested materials proved to be an enormously time-consuming and costly task. Every document, which in the end totaled almost 50,000 pieces of paper, had to be reviewed by outside legal counsel to determine whether it fell within the FTC request, whether privilege could be claimed so that it didn't have to be produced, and whether it might support in some way the case the FTC was apparently trying to make.

Additionally, the names of all individuals and firms had to be crossed out by hand in order to protect their confidentiality.

The matter dragged on until 1987, eighteen months after the FTC investigation began, when the Institute received proposed terms for ending the investigation. There was no way we could agree to their proposal, however. It would have meant a rewrite of virtually all our ethical rules, including a total lifting of our ban on contingent fees and commissions and on whom could own a CPA firm. It was the unanimous conclusion of our board of directors that capitulation on these issues would have a seriously detrimental effect on our objectivity, independence, and professionalism, and therefore on the quality of practice and on our public obligation. The board also recognized that it would be inconsistent with the views of the Council as expressed on numerous occasions.

We also knew that our plan to restructure professional standards, which the full AICPA membership was due to vote on in a matter of only a few months, would alter most of these challenged practices in ways that would accommodate most of the FTC's concerns. In fact, we were really only at odds over two major issues: our ban on contingent fees and commissions, and the form of practice in which an accounting firm could legally be organized.

By far the most contentious of these issues continued to be contingent fees and commissions. Even at this relatively early stage in what would become a six-year battle with the FTC, our outside counsel, Lou Craco, senior partner of Wilkie, Farr & Gallagher specializing in FTC and antitrust issues, predicted that "the battleground, it is fair to say, with the FTC will be about contingent fees and commissions." Craco warned Council members that if the Institute was unwilling to consent to such an order, the FTC staff was prepared to recommend to the Commission that a complaint be filed. If that happened, we would likely be forced to embark on a long and expensive litigation process. Nevertheless, at our September 1987 meeting, Council unanimously voted to reject the FTC's proposed settlement and to take the appropriate steps to resist its efforts to terminate our ethical rules. On the advice of counsel, we also suspended enforcement of the rules that had been challenged until the FTC proceeding was resolved.

Many Council members weren't too pleased with this last idea, but our outside attorneys warned us in no uncertain terms that if we continued to enforce our existing rules we would be leaving ourselves open to serious litigation from persons whom we were penalizing for performing actions that the FTC was declaring violated antitrust law. In fact, at least two members, both subjects of investigation and possible discipline by the AICPA, had already threatened litigation should the AICPA enforce its rules on commissions. There was also the thought that we ought to focus our attention on the FTC challenge, rather than on the spate of lawsuits that were certain to follow enforcement of these challenged rules.

So after changing the word "suspension" to "deferral," a motion stating that "Council reluctantly ratifies the action of the Board in deferring enforcement of the rules at this time" passed unanimously. The Council also made clear that only the enforcement of the rules was being suspended, not the rules themselves. We were not providing a window of opportunity for persons who wished to perpetrate these various acts. It wasn't an amnesty, but rather a decision not to proceed with enforcement until such time as the validity of the rule had been established.

The deferral also did not preclude individual states from taking action. With most states, the AICPA had entered into the Joint Ethics Enforcement Plan (JEEP) so that when a complaint was lodged against an AICPA member, it was sent to the appropriate state society's ethics committee for investigation. The state society then drew its own conclusions as to the member's compliance with its ethical rules, and reported that conclusion back to the Institute, seeking its concurrence. If the Institute agreed with either an administrative reprimand or referral to the Trial Board, both the society and the Institute then took action.

Under the proposed deferral, however, a complaint against a member would be sent by the AICPA to the state society. The member would be notified that the complaint had been received, that it had been docketed, and that an investigation of the complaint would be deferred until the FTC matter had been resolved. There were some states, however, that continued to enforce the rules as they had in the past.

What's more, the new professional code of ethics adopted by the full AICPA membership in January 1988 as part of the Anderson committee recommendations included a total ban of commissions and contingent fees. Exceptions were provided when the fee was fixed by courts, in most tax matters, or if determined based on the results of judicial proceedings or the findings of governmental agencies.

Six months later, at the October 1988 Council meeting, these contentious issues were again on the agenda. Once again our outside legal counsel warned us that if we failed to change our rules the profession would be opening itself up to regulation from outside forces.

"The reason this is on the agenda of Council is because in our perception this is the last chance for you to independently determine what our rules will be," explained Marvin Strait, board chairman at that time. "It is our best judgment that by the next time we meet there could well have been action that would preclude us from a rule."

But opposition was fast and vehement. "We are being asked to accept the proposition that CPAs can be impartial while accepting contingent fees for work that depends upon favorable findings," said Duane Hansen, coordinator of the California Council delegation. "This is simply incredible and will never be understood by the public."

Joe Puleo, representing the view of the board of governors of the Connecticut Society of CPAs, agreed. "On competitive bidding, uninvited solicitation, and advertising we have been told by legal counsel that we should not fight the battle, so we haven't fought any of them," said Puelo. "I suggest that even if we had spent considerable money and much effort in fighting a losing battle, what we might have gained in public recognition and pride may well have been worth the expense."

Once again Council members defeated a motion to change the AICPA's ethical rule on contingent fees, this time by one vote—98 to 97! They were continuing to hold fast against mounting pressures by taking the high road and refusing to allow contingent fees.

But the issue was not about to go away. Since the Council had repudiated the FTC's proposed consent order, throughout the remainder of 1987 the FTC staff continued to tell us that they fully intended to make an application to the full commission to proceed with an enforcement action against the AICPA. That was fine as far as the AICPA was concerned, since it was our intention to support the legality of the rules that were under challenge.

But then something unexpected happened: nothing. The threatened enforcement action was neither solicited by the staff, nor brought. Meanwhile, the enforcement of the rules continued to be in deferral. That resulted in a situation in which, by virtue of the Commission's failure to act, we had neither enforcement of the rules, nor a test of their legality.

Finally, our board of directors asked for a meeting with the FTC staff. On May 12 two Wilkie, Farr & Gallagher partners, Lou Craco and Dave Foster, along with our inside counsel, Don Schneeman, met in Washington with top FTC staff members. Our purpose in forcing the meeting was to tell the FTC staff that they had left us in limbo, that they had put a cloud over our ethical rule that they were not giving us an opportunity to dispel. "If you are going to bring a proceeding," Lou Craco told them, "do it. But if you are not going to bring a proceeding, announce that too."

As Craco related the story to Council members at our May 1988 meeting, suddenly the FTC's Director of Competition leaned forward in his chair and asked, "What would you think if we allowed a ban on contingent fees and commissions in all cases for clients for whom the attest function is performed?"

This represented an enormous change in their attitude. Their original position had been that the Sherman Act forbade tinkering with an accountant's price structure by any kind of ban on contingent fees or commissions. Then in their proposed consent order they indicated that perhaps an engagement-based ban was appropriate in the attest function only. In fact, less than a year earlier this same director of competition had addressed a letter to NASBA, the independent organization of state boards, castigating them for putting out a client-based contingent fee proposal in an exposure draft. Now he was acknowledging that contingent fees and commissions do, in fact, adversely impact our profession's independence. Suddenly the FTC staff was entertaining the notion that a client-based ban would be appropriate so that we could ban contingent fees between a CPA and any client for whom that CPA provided an attest function.

The Council quickly voted to endorse the board of directors' intention to continue discussions with the FTC, and to obtain additional information through a survey of members. A special meeting was called for August 30, 1988, to review this new information and to come to a final denouement.

This special Council meeting was an extraordinary event in our history, the first special meeting the Council had ever held. So that our outside counsel could give us the same candid legal advice they had been giving the board of directors, it was a closed meeting, which meant everyone other than Council members were asked to leave the room. Marvin Strait started off by telling a story that seemed to put our 13-year battle with the FTC in some perspective.

A ship is cruising in the ocean one night. The captain is on the bridge and sees a light in the distance. "Radio that vessel and tell them to shift their course ten degrees to the south," he orders.

The message comes back, "Shift *your* course ten degrees to the north."

Then the captain sends another message: "I'm a captain in the U.S. Navy. Shift your course ten degrees to the south."

"Well, I'm a Seaman First Class," the other voice responds. "Shift your course ten degrees to the north."

That just infuriates the Captain. "Radio that vessel and tell them we're a battleship in the U.S. Navy and they are to shift their course ten degrees to the south," he says

The message comes back: "We are a lighthouse. Shift your course ten degrees to the north."

Then Craco became serious. The question, he told us, is whether a ban on contingent fees and fully disclosed commissions for nonattest clients can be successfully defended against antitrust attack by the government or by private parties. It was his opinion that the FTC order was a highly favorable result precisely because the AICPA could *not* defend its rules as applied to a nonattest client, and that the order was very likely to be better than the result we could ever hope to obtain if a defense were attempted.

Craco then proceeded to give all 196 Council members a discourse on antitrust law. The FTC was acting pursuant to Section 5 of the Federal Trade Commission Act, he told us, which had been interpreted as being in substantial parity with various aspects of antitrust laws, including Section 1 of the Sherman Act. Antitrust law forbade any arrangement by competitors to set prices or to manipulate the methods by which prices are set. Furthermore, the Supreme Court, in cases such as *Goldfarb v. The Virginia State Bar* and in the government case against the National Society of Professional Engineers, had made it very clear that these antitrust prohibitions applied to the learned professions.

Craco explained to us that our only possible defense was the legal doctrine known as "the rule of reason." This says that an antitrust challenge can successfully be defended if the rule in question has significant benefits to the public interest. But this can only be argued if, in the words of Supreme Court Justice John Paul Stevens, "the challenged agreement is one that promotes competition."

If we decided to challenge the FTC in support of our current ban on contingent fees and commissions, our defense would have to begin with the idea that objectivity is the hallmark of our profession. It is what separates us from other professions in the eyes of the public. The protection and enhancement of this hallmark is procompetitive because it allows CPAs to compete with each other on equal terms, and because it gives CPAs a competitive advantage over other disciplines. So far so good. But this defense also required us to prove that bans on pricing methods are recognized as being important to the profession's reputation for objectivity.

There were a number of reasons why we couldn't successfully make this argument, Craco believed. First, our ability to assert that an across-the-board ban on contingent fees and disclosed commissions was uniformly recognized as necessary to the preservation of objectivity was undercut by the fact that many states had retreated from an across-the-board ban. And second, public opinion seemed to undermine our argument.

The AICPA had recently commissioned a survey by the Louis Harris organization as a way of investigating whether a rule of reason defense was feasible. The Harris survey found that most people who used our services believed that CPAs do, in fact, engage in a variety of pricing practices that seem to compromise objectivity. But those same publics still gave CPAs very high marks for objectivity. These findings clearly undercut any claim that bans on contingent fees and commissions should be preserved in nonattest relationships because the marketplace views such practices as improperly compromising our hallmark virtue of objectivity.

There was also no support in the data for the argument we would have had to make that banning contingent fees and commissions in nonattest relationships was necessary to protect perceived objectivity and independence in attest relationships. On the contrary, in making judgments about what is and what is not proper the marketplace showed a genuine ability to distinguish between audit and nonaudit services. By large margins, those surveyed believed that arrangements such as contingent fees and commissions with respect to audit clients were inappropriate. We simply would be unable to make the case that there was a spillover effect, that if you allow such practices in nonattest engagements, it will affect the public perception of whether you are objective in audit and other attest engagements.

Craco acknowledged that if that was all there was to it, we might have been compelled to take on the FTC in court just on principle. But he warned us in no uncertain terms that there was a serious threat of successful private lawsuits for treble damages based on our current rules. In fact, he told us that the chances of winning the two lawsuits that had already been threatened were much less than a 50–50 proposition.

The good news was that our extremely able negotiating team, led by Lou Craco and his colleague, David Foster, a past chairman of the ABA section on antitrust law, along with our own Don Schneeman, had been startlingly successful in their efforts to get the best deal possible. This was especially true in light of the fact that the FTC's early drafts had been totally unacceptable, leading many of us to believe that chances for a settlement were slim. In fact, the FTC had initially wanted no restrictions at all on either contingent fees or commissions. Even very late in the negotiating process, the FTC was still insisting that contingent fees could be barred in attest engagements but had to be allowed for nonattest services provided to that same client. Until May 1988, the FTC also had argued that the AICPA should be obligated to report to the FTC any state CPA society that did not conform to the order and to require as a condition of participation in JEEP that state societies agree to adhere to the provisions of the order. And finally, they were insisting that our regulation against vouching for the achievability of forecasts violated antitrust law.

The FTC was now offering us a compromise on all these points by allowing us to ban contingent fees and commissions for all attest clients, as well as all undisclosed commissions. We were only required to permit contingent fees and fully disclosed commissions for nonattest clients. The FTC staff also dropped their objection to our regulation against vouching for the achievability of forecasts after we explained to them that this was a professional standard and none of their business, and they allowed us to retain the right to recognize state sanctions and to discipline members for violating them.

Perhaps most important, the Institute was given the opportunity to submit a draft of the order, which was prepared for us by Wilkie, Farr & Gallagher in a way that provided affirmative permission for the AICPA to ban undisclosed commissions and contingent fees for attest clients. That might not seem extraordinary, but Craco assured us that it was unheard of for the FTC to issue an order allowing the target of an investigation to continue to perform acts that were the subject of the original investigation. FTC orders typically just ban what they decide to ban and remain silent about the rest. Instead, this one, if approved, would function as an insurance policy against lawsuits on antitrust grounds. That was not a result that could be obtained from a judge or jury in litigation. The order provided a basis for disposing of the threatened lawsuits that in our legal counsel's judgment we would otherwise lose. At the same time, it preserved the core values of the Institute's ethical rules and kept them safe from future attack by public or private parties.

Even the staunchest defenders of our ethical rules were comforted by our retained right to urge state CPA societies to advocate stricter regulations than the FTC order allowed. "Nothing in this Order shall prohibit the AICPA from soliciting action by any federal, state or local entity," the order stipulated. This meant the Institute could in fact tacitly support state legislation that contradicted the FTC order.

"This should not be considered as a referendum to approve commissions or contingent fees," explained Robert Petersen, a Council member from California. "Our vote today is to conclude an FTC inquiry."

Even Connecticut Council member Ben Cohen, who had been one of the most outspoken critics against any change in our prohibition against commissions and contingent fees, was won over. "At the present time my emotional and intellectual view is that we shouldn't change it, but I think logic and practicality has to enter into it," he said before calling for a vote to authorize the officers of the Institute to sign the proposed consent order. The motion passed 191 to 5.

Marvin Strait and Lou Craco concluded the meeting with a warning not to talk about the settlement, since the FTC commissioners had not yet accepted the order that their staff was recommending. But once the Council had authorized the Institute to sign the proposed order, we expected it would quickly be presented to the full commission so that the normal 60-day comment period could begin. Again, we miscalculated.

Our desire for deliberate speed in this matter was in large part due to the substantial confusion that existed concerning precisely what our rules were during this period of transition. We were fielding dozens of calls each month from members asking about the circumstances under which they could accept commissions and contingent fees. We had to explain to them that the old rules were still on the books and had to be observed, even if enforcement of them was currently being deferred.

On April 4, 1989, the approved version of the consent decree was finally published in the Federal Register, triggering the prescribed 60-day comment period. Then in July, the FTC staff forwarded its recommendations to the commission members.

But still, the FTC failed to act. Six months later, in early 1990, because of the disarray this state of affairs had produced in the enforcement of our ethical rules, Lou Craco wrote a lengthy letter to the FTC chairman laying out the situation and, as tactfully as our frustration would permit, demanding that the FTC take some action. To strengthen our position, he also organized individual meetings with each FTC member to reiterate our urgency that the matter finally be resolved. Although we were assured prompt action would be taken, more months passed and, despite a letter-writing and telephone campaign, still we heard nothing.

One reason for the delay, we later learned, was that a disagreement had developed among the FTC commissioners. Their dispute was not so much over the terms of the proposed settlement, but about whether a full record before a hearing examiner ought to be made to allow the commissioners to determine for themselves whether the consent order was appropriate. It didn't help when a new commissioner, Janet Steiger, was appointed in August 1989. She had her own agenda on consumer issues and was not about to pick a fight with the other commission members over accounting rules.

We took what political action we could, mobilizing our Washington office to lobby friendly congresspeople and senators. In our frustration, we even contemplated legal action. We couldn't compel the FTC to approve the order, but there were provisions in the Administrative Procedure Act for compelling them to take *some* action. This would hardly have been welcomed by the federal agency with whom we would have to live with in the future, but we were beginning to think it might be our only recourse.

The Council also felt that the AICPA could no longer defer enforcement of the ethical rules affected by this investigation. After all, it had been two years since the FTC had first offered us a settlement, and it was no longer prudent to continue the kind of twilight zone our deferral had caused. As a result, at its May 1990 meeting the AICPA Council, by a vote of 179–1, revoked its policy of deferring enforcement of the rules challenged by the FTC staff and directed the enforcement of all existing rules of our code of professional conduct to the full extent permitted under the proposed FTC order.

It wasn't until August 1990, more than two years after the FTC had offered us the compromise, that it finally acted on the consent order, narrowly passing it by a 3–2 vote. At last we had a ban on commissions and contingent fees that protected our independence, while at the same time protected us from legal challenge.

This saga, however, had not yet ended, as the Council continued its vigilance to make certain no other concessions were made on these issues. At its May 1991 meeting, for example, the Council tabled a recommendation from the board of directors that the AICPA advocate and support state legislative enactment of laws that allowed commissions under the same circumstances as our AICPA rules. Instead, the Council voted to have the board study the issue further and report back to the Council. At the Council's next meeting, in October 1991, the board of directors again urged state boards of accountancy to adopt the AICPA's rule on commissions. Tom

Rimerman, AICPA chairman at the time, explained that "the board was not arguing in favor of the acceptance of commissions, but that for the sake of uniformity and with a conviction that the Institute's rule adequately protected both the public and the core values of the profession, Council should encourage adoption by state boards of accountancy of the Institute's rules. If some states wanted to go further and legislate a total prohibition, the board would certainly respect that point of view but would not assist in its enactment."

But once again, the Council balked. The original motion as recommended by the board resolved:

> That the AICPA believes that the current AICPA commission rule adequately protects the public and the profession's interests. The Institute encourages the adoption of that rule by the States so as to achieve uniformity in regulation of the profession.
>
> Individual States may desire to enact more restrictive bans on commissions. AICPA will respect that decision.

First, Council member John Minert suggested the elimination of the key sentence, "The Institute encourages the adoption of that rule by the States so as to achieve uniformity in regulation of the profession." That motion passed 128-67. Then Jim Smith from Connecticut offered a substitute motion which accomplished precisely the opposite of what the board had recommended:

> WHEREAS, the AICPA Code of Professional Conduct is able to prohibit the acceptance of commissions in connection with services for clients only in certain circumstances; and
>
> WHEREAS, the Council believes that it is in the public interest that there be bylaws that prohibit the acceptance or payment of any commission by CPAs in the practice of public accountancy; and
>
> WHEREAS, such total prohibition of commission can be attained only through legislative enactment;
>
> NOW, THEREFORE, BE IT RESOLVED:
>
> That the AICPA encourages States to seek legislation to prohibit the acceptance or payment of any commission by those in the practice of public accountancy;

AND THAT, FURTHER:

The AICPA will make available its expertise and relevant materials to any State Society requesting assistance in revising the accountancy statutes of its state to include a prohibition against acceptance or payment of commissions by those engaged in the practice of public accountancy, similar to the assistance it has traditionally given in legislative efforts to achieve the goals of the Uniform Accountancy Act.

Gerry Hunt, a Council member from California, explained that when California realized that the Institute was losing the battle over commissions at the national level, it enacted a state law prohibiting CPAs in public practice from accepting commissions. "The FTC agreement was shoved down our throats, and I think to try to further shove it down by saying we need uniformity, would be a mistake. At the state level, we still should not bow and give the FTC what they wanted in the first place."

The substitute motion passed 139–59.

To some AICPA members, the loosening of our restrictions on contingent fees and commissions was considered an overdue removal of an anchor on their expanding business ventures outside their core auditing functions. Disclosed commissions and contingent fees were now allowed for recommending products or services, such as software developers selling accounting software packages or computer manufacturers selling hardware. AICPA members also could now earn commissions for selling real estate, insurance, and securities. The only restriction on commissions paid to CPAs was that they could be received only from organizations or individuals for whom the CPA did not perform attest services. To most of us, it had been evident for many years that these kinds of changes were necessary and inevitable.

To other members, however, the changes in our code represented an old and venerable profession being reduced in stature to the level of sales: a profession abandoned. Objectivity and independence were the cornerstones of what made the profession what it is, and was the reason we were held in such high regard. That perception had not come overnight, went this argument, but rather was the result of the hard work and a commitment to high ideals and strict ethics by CPAs for more than 100 years. Many Council members felt very strongly that with each loosening of our ethical rules, our status as professionals was reduced. In fact, given the revolutionary changes that have occurred to the accounting profession since the FTC Order was approved (most notably the transformation of the largest firms into global, intensely competitive conglomerates and the acquisition of many small and medium-size firms by companies like American Express and H&R Block), the warning by many Council members that these changes to our rules would jeopardize the profession's high level of professionalism must be considered somewhat prescient. It's not that the changes in the code forced members to do anything specific. It just made it easier for them to compete openly, and that, for better *and* for worse, changed our profession from a cozy little club to a sophisticated big business enterprise.

QUALITY REVIEW, PEER REVIEW

Contingent fees and commissions were not the only hotly contested rule changes we considered during these years. The second of the six Anderson committee proposals passed by the membership, making quality review mandatory for all AICPA members in public practice, also became extremely controversial within the profession. A little background here is needed to understand just how fundamental quality, or peer, review has always been to our self-regulatory efforts.

Peer reviews evaluate an accounting firm's quality control system by testing compliance of a sample of the firm's audit engagements. The results of each firm's most recent peer review then become available to the public. Each public file also includes a letter of comments, the firms' response to such a letter, and a description of any follow-up action deemed necessary by the Peer Review Committee of the SEC Practice Section (SECPS). Over time, we have found that the peer review system itself promotes quality, since firms found to have deficiencies in certain areas then have the opportunity to correct them.

"The primary thrust of the peer review process is to identify a weakness in the firms' quality control system and to insist, where appropriate, that the firm take corrective action," explained John J. McCloy at our May 1983 Council meeting. McCloy was a former advisor to several U.S. Presidents, including Presidents Roosevelt and Truman, but at the time was chairman of the POB. "Formal sanctions are to be considered only in those rare cases where the firm's quality control system cannot be relied on, and the firm itself refuses to make the corrections deemed necessary by the peer review committee," he continued.

Peer review first made headlines in the early 1970s when the SEC ordered several large firms that had been accused of audit deficiency to undergo a review of their quality controls by a team of CPAs. But even before the SEC imposed peer review as a disciplinary action, the largest accounting firms had for many years been implementing their own systems of internal inspection in order to give partners a degree of confidence that their practice was being conducted appropriately. In fact, as far back as 1963, when I was asked by Main Lafrentz & Co. to establish a national technical support office, I remember speaking with a number of large firms about what they were doing in the way of monitoring their own accounting and auditing practices in order to make certain their different offices maintained consistent standards. This was well before litigation or congressional scrutiny had become any kind of major issue for the profession.

DIVISION FOR CPA FIRMS

In order to encourage a systematic approach to peer review, in 1977 the AICPA established the Division for CPA Firms, one of the most important steps ever taken by the profession to implement its program of self-regulation. Although at this point they were still voluntary, for the first time, requirements were established to make certain that all those firms conducting attest functions would adhere to generally accepted auditing standards (GAAS). And the most important requirement—peer review every three years—would monitor adherence to these standards.

Two sections within the division for firms were created. Accounting firms that had, or expected to have, clients subject to the jurisdiction of the SEC joined the SECPS. But firms without SEC clients feared that members of the SECPS would have a competitive advantage over them, so a second unit, called the Private Companies Practice Section (PCPS), was established for those firms that audited only privately held clients. The PCPS established similar self-regulatory controls, including peer review. As a practical matter, many firms engaged in audits joined both sections.

Although the vast majority of SEC clients were audited by SECPS firms, our argument that the accounting profession could effectively police itself was weakened by the fact that membership in the SECPS was voluntary. The AICPA leadership was well aware that this had to be changed if we wanted to avoid congressional action. In fact, as the Dingell hearings (see Chapter Three) began, the Institute's leadership was quietly working behind the scenes to put to a membership vote a change in the AICPA bylaws that would require public accounting firms with SEC clients to belong to the Institute's SECPS.

In order to lay the groundwork for such a vote, in the spring of 1986 we launched a national advertising campaign extolling the virtues of hiring an accounting firm that had undergone peer review. We took considerable heat for this from some members who did not belong to the Division for CPA Firms and who therefore were not subject to peer review. Smaller firms in particular were concerned that this would give those firms that underwent peer review an advantage over those that did not. This, of course, was precisely our intention. We believed that firms participating in the peer review program were better able to do quality work than those that did not participate.

Discrimination against smaller firms has always been a concern on the part of some members. At the May 1982 Council meeting, for example, a motion was made to defer the publication of a directory of the names of the members of the Division for CPA Firms until two-thirds of the PCPS member firms had passed peer review. The motion was defeated, but a similar discussion, although this one much more intense, broke out at the May 1986 Council meeting. The debate concerned national advertisements the Division for CPA Firms had placed between October 1985 and March 1986 with the tag line, "Before you recommend a CPA to one of your clients, you should determine whether or not he is a member of the Division for Firms."

Lamar Davis, speaking on behalf of the Georgia State Society, objected to the ads because he said they "put down" CPAs who did not join the division. In response, John Abernathy, then chairman of the executive committee of the SECPS, explained that "the purpose of the ads was to educate and inform the public about the division and were not intended to denigrate any member. We don't believe they did so."

Nevertheless, Davis offered two motions, both of which were ultimately rejected by the Council. The first "resolved that this Council expresses its general disapproval of the content of the advertisements placed in several national publications by the Division for CPA Firms during the period October 1985 through March 1986."

And the second "resolved that no division, group or other segment of the AICPA membership...be permitted to advertise in a manner or style as to draw comparisons or contrasts relating to the different segments of the AICPA, which could by actual statement or implication create divisiveness and/or discourse within the membership."

In urging the Council to reject this motion, Abernathy pointed out that it "would effectively bar virtually all ads by any division that provided additional services to its members, since they could be construed as drawing comparisons and contrasts."

While the Council rejected these motions, we always tried to be sensitive to the concerns of smaller firms. After all, more than half of our AICPA membership worked in firms with less than ten members. The SECPS, for example, provided relief to smaller firms by establishing a maximum membership dues figure of $100 for firms with less than five SEC clients. It also relaxed its requirements for partner rotation and concurring partner reviews.

One of the most persuasive proponents of peer review during these years was former SEC commissioner Al Sommer, Jr., who in a number of capacities became involved with the profession's peer review efforts. In addition to being a former member of the AICPA board of directors, between 1986 and 1989 he was also chairman of the POB, which had oversight responsibility for the SECPS.

"I think firms of all sizes will find quickly that the benefits they receive far outweigh the burdens they assume," he told us at our October 1986 Council meeting. "Those benefits include the upgrading of their practice, the opportunity to receive from other auditors an objective appraisal of the quality of their service."

AN UNSUCCESSFUL ANDERSON COMMITTEE VOTE

The Anderson committee recommendation making SECPS membership mandatory, which in effect would have made peer review mandatory for all firms auditing public companies, was considered so noncontroversial that it was presented for adoption by the membership before the rest of the *Plan to Restructure* was submitted. Over the years, the SEC and the POB had repeatedly expressed the view that all CPA firms that audit publicly traded companies should belong to the SECPS and participate in its peer review program. Treadway also called for mandatory peer review for firms practicing before the SEC. Our purpose for calling for a membership vote was to demonstrate that in view of the consistent calls for firms practicing before the SEC to be subject to peer review, we could reach that goal by self-regulation rather than by government fiat.

Perhaps the rather heated discussions that had taken place at Council meetings should have been a warning to us, because despite Council's ultimately enthusiastic endorsement, only 61 percent of our membership voted in favor of the proposal, just short of the two-thirds it needed for passage. "Now the SEC is giving serious consideration to requiring peer review by government regulation for any firm practicing before it," reported board chairman Bob May at our May 1988 Council meeting. "And should the SEC not feel it has sufficient authority, the Dingell subcommittee has offered to enact legislation to require mandatory peer review."

As a result of this failed effort, the Institute undertook a comprehensive communications program to explain the six other Anderson committee recommendations that made up the *Plan to Restructure*. We had learned our lesson that a major educational program aimed at our members was needed. In this particular instance, however, our other problem was that the Anderson committee recommendations did not go far enough. They required practice monitoring for AICPA members in public practice who wished to retain their AICPA membership, which would be satisfied by participating in the peer review programs of the Division for CPA Firms, or the new quality review program established by the AICPA in cooperation with state societies that elected to participate.

But the quality review program could not be considered a substitute for mandatory SECPS membership for firms with SEC clients. Reports on quality reviews were not available to the public, so there was not the same oversight. Firms participating in the quality review program were also not subject to some SECPS requirements considered important by the SEC, including requirements for audit partner rotation and concurring partner review. The quality review program did not have an independent oversight function comparable to the SECPS Public Oversight Board, and the SEC staff was not given access to the results.

MANDATORY SECPS MEMBERSHIP

As a result, at our May 1989 meeting, the Council passed by a vote of 202 to 6 a resolution authorizing a membership ballot to require that members of the Institute shall "engage in the practice of public accounting with a firm auditing one or more SEC clients as defined by Council, only if that firm is a member of the SEC Practice Section."

We were aided during these years by our two public members, i.e., non-CPAs, of the Institute's board of directors. Ralph Saul and Kathy Wriston both brought a public consciousness to their commitment to peer review. They continually kept our eye on the ball whenever our commitment to peer review might have wavered, reminding us how important the issue was to Congress and to federal regulators.

At the end of 1989, 87 percent of AICPA members, far exceeding the two-thirds majority required for passage, voted to make SECPS membership mandatory for all firms auditing public companies. Each member of the SECPS was required to undergo a peer review every three years by another accounting firm of comparable size (or a team from several firms). A Big Six firm, in other words, would be reviewed by another Big Six firm. It was not reciprocal, however. They did not review each other. Membership in the SECPS subsequently jumped, from 519 firms in June 1989, to 1257 in August 1995.

The SECPS was also given the authority to discipline its members. In fact, it expelled some small firms that couldn't get their house in order, and on many more occasions imposed corrective measures that seemed very much like sanctions, such as more stringent continuing professional education requirements or additional peer reviews. In 1991, the first year of mandatory peer review, 83 of 300 reviews resulted in modified reports. Between 1980 and 1995, 28 cases were also referred by the SECPS to the AICPA Professional Ethics Division with a recommendation for an investigation into the work of a specific individual.

Two separate reviews were actually created by the 1989 vote. Firms with audit clients were obligated to undergo on-site reviews focused on selected audit engagements. Firms without audit clients, but which offered compilation and review services, were required to undergo off-site reviews. That meant they had to submit samples of their review and compilation reports for study. On-site reviews cost about $350 per year, and off-site reviews about $250. Firms that didn't provide audit, compilation, or review services were not subject to quality reviews, as long as they reported their exempt status annually to the AICPA. The rule was subsequently changed so that exempt firms no longer had to enroll.

The primary focus of mandatory peer review was educational and remedial. The disciplinary mechanisms built into the system were only to be used as a last resort, in exceptional circumstances such as when a member refused to cooperate. During the first few years of mandatory review some members resigned their AICPA membership or simply failed to pay their dues, but the situation soon stabilized.

To help members prepare for peer review, the Institute immediately made available a substantial amount of CPE material especially designed for various types of firms. There was no reason for small firms, for example, to prepare for a total systems review. A review of specific engagements was sufficient.

PUBLIC OVERSIGHT BOARD

The peer review process, as well as the SECPS's other activities, are overseen by the POB, created in 1977. In 1979, it was given the added responsibility of overseeing the activities of the SECPS quality control inquiry committee. That committee was charged with reviewing the implications of a firm's own procedures and standards in the light of specific lawsuits alleging defects in the audits of publicly held companies.

As originally conceived, the AICPA was to have the authority to veto members of the POB board, but the SEC insisted on total autonomy. As a result, the POB maintains its independence from both the profession and the political process by being self-perpetuating. While funded by SECPS fees, the POB elects its own board members, who set their own compensation and hire their own staff.

POB staff members participate in the field reviews of all firms with 30 or more SEC clients, as well as a sampling of about one of every five reviews of smaller members. As a further safeguard, SEC staff members randomly inspect a sample of peer review files. The POB also issues an annual report that makes public all its important actions from the previous year.

As a practical matter, there is a tremendous amount of interaction not only between the SEC and the POB, but also between the SEC and the largest accounting firms. By reviewing registration statements before they are filed, SEC staff exercise a great deal of de facto influence over the way independent auditors conduct themselves.

The first chairman of the POB was the well-respected statesman, John J. McCloy. In 1984, Arthur Wood, a retired chairman of Sears Roebuck, took over, until 1986, when former SEC Commissioner Al Sommer became chairman. Sommer, who served until 1999, immediately set out to broaden the POB's oversight responsibilities. He believed that to be effective, the POB's authority had to go beyond SECPS oversight alone.

"Peer review is not a case of auditors scratching each other's back, but rather a genuine effort to increase audit quality," Sommer explains. "But peer review cannot be viewed in isolation. It has to be seen within the overall context in which audits occur."

The AICPA leadership agreed with Al that even with an effective quality control system in place, the profession should also be concerned about any other threats to maintaining a system of quality audits. So we were enthusiastic when under his leadership the POB became more aggressive in trying to strengthen SECPS requirements.

One of the most important contributions by the POB during these years was its report entitled *In the Public Interest,* which discussed five major issues confronting the accounting profession: litigation, self-regulation, standards, public confidence, and professional practice. In all, it made twenty-five recommendations to improve financial reporting and the quality of independent audits.

The genesis of *In the Public Interest* had been a request from large firms that the POB endorse the ongoing effort to secure federal legislation that would substitute proportionate liability for the joint and several liability standard that then existed (see Chapter Two). The POB concurred that legal reform was necessary, but also declared that more needed to be done to strengthen auditing and financial reporting.

Implementation of the recommendations by the *In the Public Interest* report required action by the SECPS, the AICPA, the SEC, accounting firms, and the FASB. The Institute's government affairs committee in the Washington office was assigned overall responsibility to oversee and monitor implementation of those POB recommendations that were actionable by the AICPA, and most of the recommendations were subsequently implemented.

SECPS MEMBERSHIP

Peer review is, of course, only one of the requirements of SECPS membership, all of which affect more than 112,000 professionals at 1300 member firms that audit about 97 percent of the nation's approximately 16,000 public companies. As will be discussed in more detail in Chapter Five, all CPAs in public practice were also required to participate in at least 20 hours of CPE each year, as part of a total of at least 120 hours every three years. And professionals who devoted at least 25 percent of their time performing audits, reviews, or other attest services, or who had the overall responsibility for supervising such engagements, had to obtain at least 40 percent of their required CPE in subjects relating to accounting and auditing.

Other requirements, which were gradually strengthened during these years, included preventing a CPA from auditing the same public company indefinitely. Any CPA firm with more than five SEC audit clients and ten partners was prohibited from assigning the same partner to be in charge of an audit for more than seven consecutive years. SECPS members also had to report to the section annually the total fees received from audit clients for MAS. They were also required to report to the AICPA's Quality Control Inquiry Committee (QCIC) any litigation alleging deficiencies in the conduct of any audit they conducted. The QCIC was then authorized to direct the member firm to take corrective measures, regardless of the outcome of the particular litigation.

In order to raise a warning flag should any public company seek to change auditors as a way of obtaining a more favorable audit report, beginning in 1989 SECPS members were required to disclose to the SEC within five business days any time a corporate client had for any reason changed auditors. This "early warning system" was designed to flag potential problems at public companies before they became more serious. Requirements like these underscored the profession's ability to protect the public through self-regulation, while avoiding the direct imposition of regulation or legislation.

QUALITY CONTROL INQUIRY COMMITTEE

It is important to remember, however, that from the very beginning the role of both the Division for CPA Firms and the SECPS was corrective, rather than punitive. Their activities were designed to strengthen quality control, to prevent recurrences of problems, and to correct deficiencies in the practice of member firms. They were not intended to duplicate the oversight and enforcement responsibilities of the courts, Congress, or any regulatory agency, all of which are responsible for determining whether or not allegations of audit failures are correct.

On the other hand, since its establishment in 1979 the Institute's QCIC has made hundreds of investigations, and has proven to be a key ingredient in the Division for CPA Firms's peer review effort. The QCIC reviews possible flaws in a firm's quality control system based on allegations of negligence or incompetence. The review makes inquiries concerning the facts of the underlying litigation, but it walks a fine line so as not to jeopardize the litigation. Over the years, the QCIC has often instructed member firms to take specific measures to improve their quality control processes. In rare cases, it had to resort to sanctions. Other times it recommended changes in generally accepted auditing standards.

The role of the QCIC can be likened to that of the National Transportation Safety Board (NTSB). After an air disaster, the NTSB investigates the airline's quality control system and its compliance with certain safety standards. The NTSB is charged with protecting the public's safety, and so is the QCIC.

CONGRESS GIVES US SOME CREDIT

Finally, self-regulation had real teeth. Critics might still be able to take potshots at the details of such specific requirements as our new code of ethics, peer review, and the new educational membership rules that will be described in detail in the following chapter, but not at our commitment to make certain that all auditors of public companies adhere to stringent quality control standards.

Fortunately, our self-regulatory efforts were not going unnoticed by our most persistent critics. In the fall of 1986, Congressman Dingell had opened his subcommittee's hearings by declaring, "The public's confidence and trust in the supposedly objective and reliable information provided by accountants has been and continues to be eroded. I had originally hoped that the accounting profession could restore public trust in its integrity through vigorous self-regulation. However, self-regulation is not working and the industry has failed to earn the public trust it needs in order to be effective."

Less than two years later, Congressman Dingell had made almost a complete turnaround. "The accounting profession, through the AICPA, has made substantial improvements in its audit standards," he declared in May 1988 in his opening remarks on hearings to review the Treadway commission recommendations. "Their decisive and timely action as well as their willingness to work with the subcommittee on further improvements, is commendable."

Even more significant was the kudos we earned from two important reports by the GAO. The Dingell hearings continued sporadically until 1989, when instead of issuing its own report, the chairman asked the GAO to monitor implementation of recommendations affecting the accounting profession for the one-year period between November 1987 and November 1988. This 1989 report acknowledged that the profession had made "significant progress" in virtually all the areas under investigation and had taken a number of self-regulatory steps, which put to rest many of the criticisms that the subcommittee had aired during the previous four years.

The GAO report did indicate, however, its concern over several outstanding issues, including the reporting of fraud. Acknowledging that the accounting profession had taken a small step in this regard with its requirement that auditors report material irregularities and illegal acts directly to the audit committee, the GAO noted that it still felt that in certain circumstances, it might be appropriate to report such irregularities to officials outside the corporation. The Institute's ASB later made this suggestion mandatory.

In 1996, the GAO issued an even more comprehensive, two-volume report entitled *The Accounting Profession—Major Issues: Progress and Concerns*. Like its predecessor issued almost a decade earlier, it too had been requested by Congressman Dingell. This time it detailed the actions and progress the accounting profession had made during the previous two decades to improve accounting and auditing standards and the performance of independent audits. The report complimented the quality control programs we had implemented to ensure that professional standards were met. It also expressed its support of our self-regulatory activities, particularly those overseen by the POB.

"Most firms now have effective quality control programs to ensure adherence with professional standards," the 1996 GAO report concluded. It also commended the profession for the steps it had taken to strengthen auditor independence, such as the AICPA's revision of its code of ethics.

The 1996 report also highlighted several areas still in need of improvement. The GAO did not believe, for example, that the new standards the profession had issued to more fully define auditor responsibilities had succeeded in narrowing the expectation gap in any appreciable way. The report recommended legislation to regulate internal control procedures for all public companies, similar to what was required of banks by the Federal Deposit Insurance Corporation.

CHANGES IN EDUCATION REQUIREMENTS

The following chapter will discuss proposals four, five, and six of the *Plan to Restructure* as ratified by our membership. These concerned new educational requirements for AICPA membership. They had the most dramatic impact on most members and, therefore, in some sense on the overall profession.

Chapter 5

Education, Examination, and Experience

Accounting firms are strictly regulated, but so are individual CPAs, primarily through stringent certification requirements categorized as the three Es: education, examination, and experience. These are the key ingredients necessary for entry into the accounting profession.

To begin with, in most states CPA applicants must demonstrate to their state board of accountancy some kind of academic underpinning to their knowledge of accounting and related subjects. Increasingly, 150 semester hours of collegiate academic preparation (the equivalent of a graduate degree) education has been required, a standard provided for in the Uniform Accounting Act (see Chapter Three) and by all but a few of the fifty-four jurisdictions. All applicants must then pass the uniform CPA examination, and in most states demonstrate some practical experience. Only after fulfilling all these requirements can they become a certified public accountant, although even then they cannot simply rest on their laurels. Among other conditions, accountants must fulfill certain ongoing CPE requirements in order to keep their CPA license in good standing.

All these requirements were significantly strengthened between 1980 and 1995. They didn't come without a fight, however.

The first obstacle in the way of creating higher educational standards for CPAs was the way we are actually certified. Sometimes, of course, I wished that with the snap of our fingers the AICPA leadership could dictate the requirements for every CPA working in the United States. It certainly would have made life simpler. One hundred and fifty hours of education, a year's worth of solid work experience, pass the uniform examination, maintain your education with 120 hours of CPE over three years with a minimum of

20 hours annually, and presto—you can officially hang a CPA certificate on your office wall.

But as explained earlier, a national certification process was not to be, and nowhere did I see it on the horizon. Nevertheless, between 1980 and 1995 the accounting profession made great strides in creating more uniformity among the fifty states and four jurisdictions concerning entry into the profession. More importantly, we were able to raise the overall bar by making dramatic changes in accounting education. In fact, I believe these years witnessed a greater transformation in accounting education than in any other period in the profession's history. Most of this change was prompted by four events:

1. In 1980, the American Assembly of Collegiate Schools of Business (AACSB) for the first time adopted accreditation standards for accounting programs. Previously, the accreditation process had been limited to business programs.
2. In the spring of 1986, the American Accounting Association's (AAA) Committee on the Future Structure, Content, and Scope of Accounting Education, chaired by Norton Bedford of the University of Illinois, issued a comprehensive report warning that "accounting education as it is currently approached requires a major reorientation between now and the year 2000."
3. As a result of the Bedford report, the then Big Eight accounting firms formed the Accounting Education Change Commission (AECC), which subsequently awarded more than $4 million in grants to schools to invest in innovative curriculum.
4. In 1988, as part of the Anderson committee's recommended *Plan to Restructure*, the AICPA adopted a membership admission requirement that after the year 2000, applicants must have at least 150 college-level semester hours of accounting education and must fulfill certain CPE requirements.

Let's now look at each of these events in greater detail.

AACSB ACCREDITATION REQUIREMENTS

Founded in 1916, the AACSB began to accredit business schools with the adoption of its first standards in 1919. But it wasn't until 1980 that the AACSB started to accredit accounting programs.

This accreditation process became a joint effort by the AACSB and the accounting profession. Included on the AACSB accreditation committee were representatives of the AICPA, the AAA, the National Association of Accountants, and the Financial Executives Institute. Moreover, the profession, primarily through the AICPA, agreed to provide financial support to the new accreditation program.

Accreditation is to the educational institution as peer or quality review is to the CPA. An accounting department invites a team of outsiders to visit the campus and to examine in detail the educational process in order to make certain it meets essential standards. The visitation team meets separately with students, faculty, administrators, alumni, and practicing accountants. They look at admission requirements, curriculum, computer and library facilities, teaching loads, and class size, among other factors.

Most of the accounting accreditation standards adopted in 1980 were "input standards," concerned with making certain that accounting faculty had an appropriate level of professional certification, experience, and academic credentials. One of the more important initial requirements was that a significant percentage of faculty members had to have "recent relevant experience" in accounting. This requirement was in response to a concern on the part of many of us that CPA faculty lacked sufficient practical experience.

One method many universities used to fulfill this requirement was to hire as part-time instructors CPAs actively involved in public practice. This proved to be an effective way to keep students abreast of the latest areas of accounting expertise. Over the years, however, we continued to be concerned about the shift of academic influence from the CPA practitioner to the academic community. At our October 1995 Council meeting, for example, Tom Nelson, former president of the Utah Association of CPAs and the AICPA/NASBA Task Force on the 150-hour Educational Requirement, spoke about his own experience on accepting our Outstanding Educator Award.

"When I first entered the profession most accounting educators were CPAs," he told us. "Very few had accounting Ph.D.s. But today it seems that the only acceptable credential is the Ph.D., while the CPA means little. In fact, at some institutions it is even regarded as a negative. As a result, many new hires have little, if any, practical experience. To make matters worse, they are loaded with research responsibilities so that they have no opportunity to interact with the practicing profession."

The Outstanding Educator Award, instituted before I became AICPA president, is given annually to a member of the academic community who has done something special to encourage accounting students. It is awarded in recognition that the future of the accounting profession is directly related to the quality of young people entering the profession, and that the competence of those entering the profession depends to a very great extent on how well they are prepared by their college and postgraduate education. The award has since been renamed the Distinguished Achievement in Accounting Education Award.

By 1983, the AACSB had accredited the accounting programs at 28 universities. By 1999, the number neared 150. As a further sign of our new stature, the executive committee of the AACSB approved a revamping of the organizational structure of accounting accreditation committees. A new group, the Accounting Standards Committee, was charged with reviewing accounting accreditation standards and with proposing any changes to the AACSB.

In April 1989, this committee, with the support of the AACSB, began a two-year review aimed at redefining the goals of the accreditation process and revising the standards and procedures used to meet them. This accreditation project issued new "mission-based" criteria, which were soon approved by the AACSB. They were aimed at offering schools the opportunity to be judged by how well they fulfilled their individual mission. The new standards were a recognition on the part of the AACSB that quality education could be delivered in a number of different ways, all of which could be worthwhile and commendable. This revised philosophy reaffirmed our belief that there was no single correct way to approach accounting education.

A school still had to meet certain standards, of course. It had to offer a baccalaureate degree as a foundation of knowledge, and have a clear educational mission. As Milton R. Blood, the AACSB's director of accreditation, explained, "We're trying to build a process that encourages innovation. Unfortunately, the old standards sometimes were perceived as Procrustean. We want to move away from that."

The new standards kept specific curriculum requirements to the minimum so that accounting education programs could take advantage of their own unique academic strengths. They also provided a more detailed interpretation of the qualifications required of faculty. Schools could now focus faculty efforts toward their own choice of instructional development, applied scholarship, or basic scholarship.

ASSOCIATION OF COLLEGIATE BUSINESS SCHOOLS AND PROGRAMS

The AACSB's new flexibility regarding standards was at least in part a response to a rival accreditation authority created in 1989. Many schools, particularly two-year junior colleges, had become dissatisfied with the current system. They were concerned that the AACSB, with its emphasis on research, discriminated against the many quality schools that had teaching excellence as their primary objective. The AACSB also ignored two-year colleges altogether.

As a result, in 1988 representatives from 150 business programs met to discuss alternatives to the existing accreditation system. The following year, they formed the Association of Collegiate Business Schools and Programs (ACBSP), with a mission to "establish, promote, and recognize educational standards that contribute to the continuous improvement of business education."

Although accreditation by the ACBSP was not considered as prestigious as that given by the AACSB, the AICPA immediately accepted its authority. We believed there was more than enough room for the two services to coexist effectively. The ACBSP's requirements might be more relaxed than those of the AACSB, particularly with regard to the mandated level of technical facilities. But the ACBSP was nevertheless a worthwhile effort to legitimize the many quality educational institutions that either offered two-year programs only, or did not have the adequate endowment to make a major investment in research. All told, by 1995 fifteen to twenty schools of accountancy and about eighty other schools had qualified for accounting accreditation.

BEDFORD REPORT

Accreditation was, of course, only one step on the road to improving accounting education. By 1980, there was a growing realization within the profession that something radical had to be done to improve the accounting curriculum. The profession was rapidly entering areas of business never dreamt of by most CPAs even a few years earlier, and accounting education was having difficulty meeting the needs of an expanding profession.

One catalyst for change proved to be a committee appointed in 1987 by the AAA to investigate the future structure, content, and scope of accounting education. Chaired by Norton M. Bedford, a professor of accounting at the University of Illinois and a former president of the AAA, the committee began by studying the evolution of the accounting profession between 1925 and 1980, on the one hand, and developments in accounting education during that same period, on the other. It quickly became apparent to the committee members that while the responsibilities and requirements of the profession had changed dramatically, the substance of accounting education had remained essentially the same.

The Bedford report summarized its findings with twenty-five specific recommendations, to be implemented during a fifteen-year transitional period between 1985 and 2000. Taken together, they suggested that university accounting education should be restructured into a three-phase program of general education, general professional accounting education, and specialized accounting education.

The first phase, representing the first two years of college, would provide future accountants the fundamentals of a humanities and arts and sciences education, and make certain they understood the implications of both the global economy and new technologies. During the second two years, students would be taught the basics of accounting necessary for entry into the profession. The fifth year of study would provide them with an in depth understanding of one or more specific areas, such as auditing, information systems, public reporting, taxation, MAS, or international accounting.

The committee also recognized that in addition to technical skills, new CPAs had to be prepared for a lifetime of learning. That only could come by knowing how to reason logically, think creatively, and solve problems. Accounting education in the 1980s had to be built on a broad base of general education, including the humanities, arts, and sciences, the committee concluded. Effective communications and interpersonal relationship skills could no longer be ignored.

The Bedford report also concluded that a CPA's education should include a historical perspective of the accountant's evolving role in society, as well as an appreciation of the profession's ethical and professional responsibilities. "The personal capacities of students to interact well with others, assume responsibilities, reason logically, think creatively, appreciate ethical standards and conduct, and communicate effectively must be enhanced," Norton Bedford wrote in the August 1987 issue of the *Journal of Accountancy*. "To do this, accounting educators need to emphasize student learning rather than delivering lectures and presenting problem demonstrations."

AICPA VICE PRESIDENT FOR EDUCATION

At the same time the academic community was beginning to make changes from within, the AICPA was recognizing the need to be more active in influencing the conduct and quality of accounting education. As a catalyst, in February 1988 we created a new position at the Institute, vice president of education, to help formulate a program that would enhance accounting education. Rick Elam, a former professor at the University of Missouri Columbia and past president of the Federation of Schools of Accounting, was subsequently engaged to fill the new position.

By 1992, Rick had helped initiate several AICPA programs, with particular emphasis on improving teaching effectiveness. We began by cosponsoring, along with Georgia State University, our first conference for accounting educators. This was part of an ongoing effort to encourage innovation. We also initiated the professional/practitioner case development program designed to generate a set of "real world" cases for use in financial and managerial accounting courses.

ACCOUNTING EDUCATION CHANGE COMMISSION

In 1989, the then Big Eight accounting firms issued their own report, entitled *Perspectives on Education: Capabilities for Success in the Accounting Profession*, endorsing the major conclusions of the Bedford committee. "A growing gap exists," it concluded, "between what accountants do and what accounting educators teach."

The *Perspectives* report concluded that new CPAs should develop a wide range of professional skills over and above the traditional accounting curriculum, including interpersonal communications and general intellectual skills. The report also observed that many accounting educators "question the effectiveness of traditional teaching and learning methods," and are concerned that many accounting graduates "do not know how to communicate, cannot reason logically, and have limited problem-solving abilities." The report cited all these factors as "evidence of the need to fundamentally change university accounting educational processes."

The same Big Eight firms that issued the *Perspectives* report also wanted to make certain that their efforts resulted in something more than just theoretical, high-sounding recommendations. Toward that end, they entered into a partnership with the AAA to form the Accounting Education Change Commission (AECC). Its task was to provide leadership in the effort to make accounting education responsive to the needs of those entering a variety of career paths. In order to ensure a broad perspective, commission members included representatives not only from the AICPA and AAA but also regulators, NASBA, and business school deans (represented by the AACSB). The AECC then proceeded to make $4 million worth of grants available to encourage innovative accounting programs that adopted the recommendations of the Bedford committee and other advocates for change.

In subsequent years, the AECC proceeded to publish a series of issues statements, thus continuing to be a catalyst for improving the academic preparation of accountants. The commission's entire objective was to help create an environment in which entrants to the accounting profession would possess the skills, knowledge, and attitudes required for success in the many different accounting careers that were suddenly possible.

Although each university interpreted the recommendations from the Bedford report, *Perspectives on Education*, and the AECC differently, there was much common ground. Table 5.1, reprinted from the August 1993 issue of the *Journal of Accountancy*, presents some key features of the new approach as described by then AECC Chairman Doyle Z. Williams, a professor of accounting at the University of Southern California and a past president of the AAA.

While much was accomplished, transforming an entire educational system proved to be a formidable task. A 1994 survey conducted by the Institute of Management Accountants and the Financial Executives Institute of 2700 corporate executives from a broad cross section of industries concluded that

Table 5.1. Features of New versus Traditional Approach

Traditional Approach	New Approach
Heavy emphasis on technical courses in accounting	Broader emphasis on general education and business and organizational knowledge
Little integration of subject matter. Accounting courses taught in isolation	Heavy integration of tax, managerial accounting, financial accounting, systems and auditing
Heavy emphasis on calculating one right answer	Increased emphasis on solving unstructured problems such as use of cases
Heavy emphasis on teaching rules	Increased emphasis on the learning process, on learning to learn
Heavy emphasis on teaching to the Uniform CPA Examination	Recognition of a broader objective
Little attention to communications and interpersonal skills	Increased emphasis throughout accounting curriculum on writing, presentation, and interpersonal skills
Students as passive recipients of knowledge	Students as active participants in learning
Technology used sparingly in non-computer courses	Use of technology integrated throughout accounting curriculum
Introductory accounting focused on preparing external financial reports, journal entries, postings, etc.	Introductory accounting focused on role of accounting in society and in organizations; increased focus on using accounting information for decision making

university accounting programs were still not emphasizing the appropriate accounting knowledge and skills areas. The study found that graduates were underprepared in those areas rated most important by corporate executives, such as budgeting, product costing, and asset management. In less important areas, such as personal income taxes and external auditing, they were overprepared. One way we were able to standardize an upgrade in accounting education was by implementing a national standard of 150 hours of college education.

150 HOURS OF COLLEGE-LEVEL EDUCATION

The CPA profession was one of only a few major professions (engineering comes to mind) that did not require a postgraduate education. Instead, we at least wanted to mandate some kind of additional education. Toward that end, we initiated efforts aimed at requiring all CPA candidates to have 150 hours of college-level education. In practice, that meant between 20 and 30 hours of graduate-level training, in addition to the usual 120 hours or so of undergraduate study. In most instances that meant a graduate degree.

As far back as 1959, the AICPA had stressed the importance of a postbaccalaureate education for all accountants. In 1969, the Institute formally went on record that in order to obtain the requisite body of knowledge to become a CPA, five years of college study was needed. Then in 1978, the Institute for the first time officially endorsed the idea of 150 semester hours as the minimum education needed to sit for the CPA examination. It also issued a report—*Education Requirement for Entry Into the Accounting Profession*—that named general education, general business education, and accounting education as the three essential ingredients of a quality accounting education.

The first important effort to make the 150-hour rule mandatory was our establishment in 1981 of the Commission of Professional Accounting Education. Chaired by Wayne J. Albers, a former partner at Ernst & Whinney, the commission was charged with studying the benefits of a five-year education requirement for CPAs and developing plans for implementing it nationwide. It also was asked to investigate why the 150-hour requirement had been so slow to be adopted at the state level. Eighteen months later, the commission issued its recommendations.

Two reports were actually released by the commission in August 1983. The first, *A Post-Baccalaureate Education Requirement for the CPA Profession*, presented evidence supporting a requirement for a fifth year of study. The second, *Implementation of a Post-Baccalaureate Education for the CPA Profession*, presented strategies for eliciting professionwide support for this requirement and for achieving enactment of state legislation. Later, acting on a joint recommendation of the Institute's education and state legislative committees, we authorized the appointment of a special committee to develop a program for legislative enactment of the postbaccalaureate requirement for entry into the profession.

In 1984, the 150-hour requirement was a main feature of the new Model Public Accountancy Bill, the joint effort of the AICPA and NASBA to have states adopt uniform certification standards (see Chapter Three). The bill stated that the education requirement for entering CPAs should be "a baccalaureate degree or its equivalent conferred by a college or university acceptable to the (state) board and not less than 30 semester hours of additional study, the total educational program to include an accounting concentration or equivalent."

The AECC, of course, supported our cause. In August 1993 AECC chairman Doyle Williams wrote, "The 150-hour requirement for future CPAs provides an excellent opportunity for accounting faculties to reexamine the entire curriculum and should result in fundamental changes in the structure and delivery of accounting education from the first course to the last."

Some segments of the profession, however, seemed, at best, to be lukewarm to the 150-hour requirement. Both the AAA and the AACSB, for example, did not support the 150-hour provision in the model bill. We frankly didn't understand this sentiment. In an attempt to investigate their position, in early 1984 the Institute appointed a special committee on postbaccalaureate education requirement, chaired by Robert C. Ellyson, managing partner of the Miami office of Coopers & Lybrand and an AICPA board member.

One of the special committee's first actions was to study the various opinion polls that had been taken on the topic. It found that the rank and file generally favored an additional education requirement. According to a 1984 survey, 65 percent of Texas CPAs in public practice supported a five-year education requirement. Another survey found that 70 percent of Florida state society members supported the language found in the model bill.

But the investigation by the Institute's special committee also discovered that support was less enthusiastic among the largest accounting firms. This was primarily due to two factors. First, the large firms were more likely to have the wherewithal to support in-house training programs of their own, which made additional schooling less important to them. And second, these firms were concerned that more educated recruits would expect higher starting salaries, thereby driving up their personnel costs. But as Gary Previts noted in the October 1991 issue of the *Journal of Accountancy*, "If you think the cost of education is high, consider the cost of incompetence."

Some large firms also appeared concerned that making it more difficult to become a CPA by adding additional educational requirements could decrease the pool of qualified accountants, thereby somehow making it more difficult to recruit the best and brightest young people to the profession.

We tried to make the case to our members in public practice that in the face of new specialties, new technologies, and the growing complexity of our global economy, it was more important than ever to ensure that future AICPA members met the highest standards of competence and professionalism. As ammunition, we used the fact that while the majority of accounting students hired by the large firms had bachelor's degrees only, the majority of new partners had master's degrees. That seemed to us to mean that it was personally advantageous for the individual CPA to obtain a thorough, quality accounting education.

AACSB members (who were primarily business school deans) seemed to be concerned that a movement toward professional schools of accountancy could become independent of the business schools. We tried to set their minds to rest that this was no one's hidden agenda.

Some other studies gave us ammunition in our efforts to promote the 150-hour requirement. In one, Philip H. Siegel, an associate accounting professor at San Francisco State University, concluded in 1987 that a postbaccalaureate education, coupled with an undergraduate major in accounting, was a superior educational preparation for auditors in public accounting firms compared with a bachelor's degree alone. Siegel based his finding on an analysis of auditing professionals hired within the previous five years by five international CPA firms. He focused on three measurements of performance: the annual evaluation given auditors by their supervisors, how long it took for an auditor to be promoted, and turnover rates. The data demonstrated that individuals with master's degrees outperformed those with only bachelor's degrees in all three areas.

Also helping our cause was the situation in Florida, which in 1981 became the first state to mandate a fifth year of accounting education. The immediate result was an alarming drop in the number of first-time candidates taking the CPA exam in Florida, which plummeted from 2306 in November 1983 to only 12 in May 1984. But the drop-off was short-lived, indicating that it was obviously due to a surge in the number of candidates who had taken the exam in the preceding year or two as a way of beating the deadline for the five-year requirement. Within several years, the number of candidates rebounded to pre-1981 levels.

Even more supportive of our argument was the fact that the quality of Florida graduates improved, as measured by the pass rate for candidates taking the CPA exam for the first time. "Staff accountants with a fifth year of education are more knowledgeable and quicker studies; they deal better with complex accounting and tax issues, grasp new concepts quickly, and have a sounder background in communications, management sciences, logic, finance, and marketing," wrote John K. Simmons, a past AAA president and a professor of accounting at the University of Florida, in the October 1991 *Journal of Accountancy*.

The Florida experience also helped put to rest concern on the part of some of the larger firms that a fifth year of education would increase salary costs. While starting salaries did rise in Florida, higher billings rates mitigated the increases since the new hires were more qualified. Also, because more candidates passed the CPA exam on the first try, firms incurred less costs in days lost while employees were preparing to retake the exam.

The final validation of Florida's pioneering effort occurred in 1984. After some Florida CPAs requested its repeal, members of the state accountancy board voted by an 80 percent margin to continue the fifth-year rule.

Taken together, these activities by the AICPA and others were part of a decades-long effort to lobby our various constituencies and to lay the foundation for a membership vote to make 150 hours of education a criterion for AICPA membership. The final catalyst toward this end was, of course, the recommendation by the Anderson committee, which led to the 150-hour requirement being part of the *Plan to Restructure Professional Standards*. By an 82 percent to 18 percent margin, the membership agreed that anyone seeking AICPA membership after the year 2000 must have 150 semester hours of education, including a bachelor's degree, from an accredited college or university.

Even after the membership vote, however, we had to remain vigilant in our defense of the 150-hour requirement. One particularly annoying criticism came in May 1995, in a report by the American Legislative Exchange Council (ALEC), an educational association of state legislators. ALEC's description of the 150-hour requirement as an unnecessary burden on CPA candidates was rife with false predictions and outright inaccuracies, and we at the Institute felt it should not go unanswered. The report warned, for example, that the 150-hour rule would precipitate a drop in the number of people wanting to study accounting. But in the five states that had already implemented the requirement, schools had actually been forced to raise entrance requirements to offset the growth of applicants. The ALEC paper also declared that the requirement put many small firms at risk, a claim that had no basis in fact.

In response, NASBA President David A. Costello and I sent a sharply worded letter to ALEC's executive director. "As an association of state legislators who represent the public, you have a duty to get the facts straight before publishing a position paper for public dissemination," we told him, before refuting ALEC's various unfounded theses point by point. Our arguments were buttressed by the fact that by 1995, most states had adopted the 150-hour requirement with wide bipartisan support.

CONTINUING PROFESSIONAL EDUCATION

As explained in the previous chapter, three of the six Anderson proposals as submitted to a membership vote in late 1987 and early 1988 concerned education requirements. As just discussed, one of them was 150 collegiate semester hours of post high school education. The other two concerned CPE.

The AICPA has been involved in promoting CPE to its members since 1958. In fact, in 1961 I worked for the Institute in what was then called the professional development division as administrator of a training program for newly hired staff accountants. But it wasn't until 1981 that the concept of a truly national CPE curriculum began to take shape. This was long overdue, since for a profession such as accounting, which is forever expanding and adapting to the global business environment, keeping abreast of new developments is critical.

One of the steps we took to reinforce the importance of CPE was to elevate the person responsible for its activities to a new executive staff position, vice president of CPE. Thomas Murphy, who had been in charge of the Institute's publishing activities, was the first person to fill the post. Previously, the AICPA's CPE efforts had been administered by accounting education professionals, but we felt that someone with more of a business orientation stood a better chance of maintaining the program's financial viability. Over the years, this proved to be a difficult goal, as a slew of low-cost competition emerged to capture CPE revenue. We were hindered with a responsibility that third-party providers of CPE did not have, namely, to ensure the availability of adequate programming for CPAs in a wide range of subjects. We embarked on an extensive effort to make the Institute's CPE programs more relevant to the expertise our members needed in their various areas of employment.

One CPE milestone in this regard was the Institute's national curriculum project called *A Pathway to Excellence.* Initiated in 1980, this five-year effort culminated in a report that divided our new CPE curriculum into six fields of study:

- Accounting and auditing
- Consulting
- Management
- Personal development
- Specialized knowledge and applications
- Taxation

We felt that only after CPE programs were available in virtually every conceivable field, could we pursue our goal of making CPE mandatory for all AICPA members. Toward that end, we appointed a special committee, which in September 1985 issued a rationale for mandatory CPE organized into three principal arguments:

- The public interest is served by mandatory CPE for CPAs.
- CPE is needed to maintain professional competency.
- CPE should be mandatory rather than voluntary.

"Mandatory CPE provides the general public with a measure of assurance that CPAs are maintaining their competency by staying abreast of newly issued accounting and auditing standards, changes in the tax code and regulations, and other governmental requirements affecting matters dealt with or handled by CPAs," the report concluded.

In order to lay the foundation for making CPE mandatory, we had to dispel a number of misconceptions that had arisen over the years, including that it was not cost-effective and that CPAs who worked outside large metropolitan areas would have a difficult time finding the appropriate CPE courses. The special committee also helped make the case that neither peer reviews nor reexamination were acceptable substitutes for mandatory CPE. Peer reviews do not impart knowledge of new technical or regulatory rules or regulations, and the CPA exam tests one's knowledge on a broad range of subjects, which is inappropriate for the experienced practitioner who has spent his or her career specializing in one concentrated field of expertise.

We also understood that mandatory CPE had to be flexible in order to allow each individual to choose the type of course that made the most sense for his or her particular practice. We believed that members themselves were in the best position to determine the CPE courses that would contribute to their technical competence and allow them to best serve the needs of their clients.

Just as with the 150-hour requirement, the seminal event in CPE during these years was the Anderson committee recommendations. Proposals four and five of the *Plan to Restructure*, mandating CPE for members in public practice, generated relatively little debate. By 1987, forty-seven states had some kind of CPE requirements anyway, so the new rule represented little change for the vast majority of members.

The requirement for those not in public practice generated much more controversy. In the end, however, the argument that the standards of AICPA members in private practice (those working in industry, government, and education) should be elevated to meet those working in public practice (in firms acting as independent public auditors) carried the day.

Thanks to an intensive lobbying effort by Marvin Strait and others (see Chapter Four), mandatory CPE for AICPA members in public practice passed with 90 percent of the vote, and for members not in public practice with 74 percent. Members in public practice were required to have 120 hours of CPE over a three-year period (with a minimum of 20 hours per year), while 60 hours (with a minimum of 10 hours per year) was mandated for members in private practice.

During subsequent years, we had to remain vigilant in order to withstand various challenges or efforts to dilute the effects of mandatory CPE. At the September 1989 Council meeting, for example, a motion was sponsored by the Society of Louisiana CPAs to eliminate the 20-hour yearly minimum CPE requirement. George Anderson spoke against the suggestion, arguing that members would be tempted to obtain all the required 120 hours in the third year, which would be inefficient. Others spoke against the proposal, including Marvin Strait and Texas CPA Society President Claude Wilson, who told Council members that "it sends the wrong message to the public at a time when the profession is under attack from numerous sources because it could be perceived as backing off of our CPE requirement." The motion was overwhelmingly rejected.

As CPE became mandatory, we also tried to make it clear that the new requirements were for AICPA membership only, and were not the same as CPE requirements set by state boards or state societies, which were often even more stringent. At the time, for example, eleven states required a minimum number of hours of CPE in accounting and auditing. Two states required a minimum number of hours in taxation.

In 1980, the AICPA was the major developer of CPE courses. Over the years, the states developed their own courses, and then began obtaining them from lower cost third parties. One of my regrets as president was that the Institute was unable to continue as the principal developer of high-quality programs. But we weren't able to compete on price with low-cost third-party vendors, and in the end, marketplace forces caused most of the CPE business to become localized and dominated by numerous third-party vendors.

As with most consumer products and services, however, competition did serve the purpose of making CPE programs widely available and at a competitive cost. The number of CPE courses dramatically increased, making them much more affordable and accessible to the individual AICPA member. That was our goal in the first place. By 1995, in California alone state society members could choose from more than sixty-five courses aimed at industry members and more than a dozen aimed at those working in government. In addition to the main CPA specialties such as taxes, financial management, and advisory services, industry members were offered courses on such diverse topics as real estate, entertainment, agriculture, and the wine industry.

The Institute did continue to play a vital role in making certain that the appropriate CPE opportunities were available that would ensure the professional competence of every member of the accounting profession. We helped make CPE more accessible to members and to expand the available subject matter. We also pioneered our own low-cost alternatives, such as the *CPE Direct* self-study program introduced in 1993 that offered 24 hours of CPE credit for reading twelve issues of the *Journal of Accountancy* and reviewing quarterly study guides that fleshed out material covered in the magazine. A similar program was created for reading *The Tax Adviser*. We also made the AICPA's CPE requirements broad enough to embrace a wide variety of programs, including conferences at local universities, seminars offered by state CPA societies, self-study programs, in-house courses offered by a company or CPA firm, university or graduate courses, or even teaching a formal CPE program, as long as it increased the CPA's professional competence. By allowing wide latitude in the kind of continuing education members were allowed to use, we made it easy for Institute members to select topics of study that would enhance their professional competence and keep them in the forefront of their chosen fields.

Over the years, a great deal of effort was expended by the AICPA leadership to improve the effectiveness and efficiency of our CPE activity. During the mid-1980s, as a result of strong lobbying by a few state societies, we came close to moving the CPE division to a more central location in the Midwest. But the most comprehensive study we undertook was led by Terry Sanford, former U.S. senator and governor of North Carolina and president of Duke University. In 1993, we asked him to study our CPE organization and recommend ways to improve it.

The Sanford committee observed that in a dynamic, rapidly changing competitive marketplace, the tensions and conflicts that existed between the Institute and some state societies had to be put aside for the good of the profession. The committee declared that a new boldness was needed to come up with a new structure for implementing CPE. "My personal view is that you wanted us to offer you tangible recommendations, not just to remind you of the problems you already know exist," Governor Sanford told us at our May 1993 Council meeting.

The Sanford committee offered nine specific recommendations:

1. Creation of a new organization, called the Alliance for Learning, that would take over from the Institute's CPE executive committee overall responsibility for all AICPA and state society CPE efforts
2. Establishment of a cooperative arrangement between the AICPA and other professional associations and commercial providers
3. Elimination of artificial barriers to the marketplace
4. Maintenance of CPE as a self-supporting service, although with the recognition that certain necessary courses may not always be profitable
5. Study, review, and revision of all current CPE products
6. Reevaluation of the AICPA *Statement on Standards for Formal CPE Program*, taking into account a focus on both formal and nonformal learning
7. Development of a business plan based on market research
8. Utilization of new technologies
9. Establishment of a national database for the ordering and delivery of all CPE and other products

Later that same year, Jake Netterville, who subsequently served a term as AICPA chairman, was appointed to head a special task force to evaluate the

Sanford committee's recommendations. The task force concluded that eight of the Sanford committee's nine recommendations should be implemented immediately, but rejected the committee's centerpiece proposal, the establishment of the Alliance for Learning. On paper, the idea of creating one organization that would share state and national resources and thereby meet the educational and information needs of all our members was a good one. But as a practical matter, the business of CPE was the lifeblood of the largest and most influential state societies, particularly the "Big Eight" of California, Texas, New York, Ohio, Illinois, Florida, Michigan, and Pennsylvania. From their point of view, it made no financial sense to centralize their CPE activities within the AICPA.

Instead, a CPE board of management was created to replace the existing CPE executive committee. Marvin Strait was appointed as its first chairman. Proposals continued to fly, however, on ways in which we could spur more cooperation between AICPA and state CPE efforts. At the May 1995 Council meeting, Marvin decried the rivalry between the AICPA and the state societies. "It makes no sense that our organizations, which essentially have the same members, are competing with one another," he told us. "Many times I have difficulty determining how members decide whom they are representing, because practically everyone in this room is a member of both a state Society and the AICPA. What's needed is more than just fine tuning, but a major fundamental change in the entire relationship."

Marvin announced the formation of a planning group made up of four society executives, four CPE Directors, and the CPE board of management, as well as both me and my successor, Barry Melancon. As a facilitator, Marvin recruited Charles Peck, a former vice president of Simon & Schuster, with twenty-five years experience in selling and developing educational materials. "The question we have to ask ourselves," Marvin told us, "is, 'If we were starting all over, what kind of a system would we develop?' We have to ask members to put their respective organizations' interests secondary to the interests of the profession. There must be no sacred cows; it's all on the table. We need to develop a new model so that the Institute and the state societies can together deliver CPE in the most effective manner possible."

EXAMINATION

The great equalizer in the certification process has always been the uniform CPA exam, developed and graded by the AICPA and administered twice per year in each state. The exam was first made available in June 1917 by the AICPA's precursor organization. All CPA candidates, regardless of the state where they reside, must pass the exam in order to qualify for their CPA certificate and be permitted to practice. It is a rigorous test of an applicant's knowledge, so rigorous in fact that in 1995 only 17 percent of those who took the exam for the first time passed all four sections. But those expecting to take the exam one day should take heart. The pass rate improves as candidates retake it.

Some states, either because of financial limitations in state budgets, or for the sake of convenience, do not want to take on the responsibility of administering the uniform exam. So the AICPA and NASBA jointly established the Examination Services Corporation (ESCORP) to administer the exam in any state requesting its assistance. By 1986, ESCORP had become financially viable and the Institute was able to withdraw its financial support from the venture.

While it is true that all jurisdictions require candidates to pass the uniform CPA exam, there are differences in the minimum grade necessary to pass it and the time a candidate must wait before retaking a section he or she has failed. Also, as of 1995, thirty-five states required an additional exam or qualifying course in professional ethics.

In 1991, we began to institute some significant changes in the uniform CPA exam to make it more relevant to the diverse employment opportunities open to those who passed it. The AICPA Board of Examiners (the Institute's committee responsible for the exam) mandated that beginning in May 1994, the exam would be restructured into four new sections.

Many of the changes were based on a 1991 survey of the professional activities of 1800 CPAs. The results were published in a report called *Practice Analysis of Certified Public Accountants in Public Practices* and provided the basis to ensure that the exam tested the knowledge and skills relevant to the practice of CPAs within rapidly changing business, government, and financial environments.

In 1991, the board of examiners also declared that effective May 1996, the exam would be nondisclosed, meaning that the questions and answers no longer would be made public after the exam has been administered. Then in 1994, the examination was reduced from two and a half to two days. Other changes included:

- The four sections of the exam were more efficiently organized. The accounting practice and theory sections, for example, were redistributed into two new accounting and reporting sections. Some of the questions on professional responsibilities were shifted from the auditing section to the business law and professional responsibilities section.

- A greater emphasis was placed on auditing government and not-for-profit organizations.
- A new emphasis on writing skills was added.
- Candidates were allowed to use calculators on the financial accounting and reporting sections.
- For the first time in twenty years, questions not in multiple choice form were included.
- A new section on professional duties and responsibilities was included to emphasize that the ethical aspects of practice are as important as technical competency.

The final initiative during these years was the issuance by the board of examiners of the *Invitation to Comment—Conversion of the Uniform CPA Examination to a Computer-Based Examination*. The purpose of this initiative was fourfold:

- To inform boards of accountancy and other interested parties of the board of examiners' intention to convert the uniform CPA examination to a computer-based exam
- To provide information about the various types of computer-based tests and their benefits and costs
- To provide examples of several computer-based test models and how they might be implemented with the uniform CPA exam
- To obtain the views of boards of accountancy and other interested parties on the feasibility and acceptability of a computer-based exam and on the nature of that exam

Responses from boards of accountancy, NASBA, and others indicated that the majority of CPAs favored computerization, although there was not a clear consensus on a single computerization model. Among the concerns expressed were cost, security issues, and that a computerized exam could adequately test communications skills. Subsequent to my retirement, all these concerns were explored by an AICPA/NASBA computerization implementation committee. The first computerized exam is scheduled for the year 2003.

EXPERIENCE

There has always been criticism that even after 150 hours of education, graduating students are not prepared for the rigors and the broad-based knowledge that most employment as an accountant demands. Other professions are no different, of course. How many graduating law students are really prepared to go to trial the day after they pass the bar exam? And who would be eager to be operated on by a medical school graduate before he or she has gone through a rigorous internship program?

In the accounting profession, most states require a certain level of employment experience before certification is allowed. But these requirements have generally been relaxed in recent years, mainly because there has never been a set of ground rules concerning precisely what this experience should entail. Typically, the only requirement is that the applicant obtain a letter from another CPA certifying that he or she has been employed for the requisite period of time. When I first went to work in New York State, I needed three years of experience before I could complete the exam process and become a CPA. Now, as part of the Uniform Accountancy Act (UAA), only one year is required.

Experience requirements vary widely among the fifty states, with more than half granting a CPA certificate without requiring any practical work experience, and the rest requiring anywhere from one to five years. New York probably has the most unusual requirement, or lack thereof. If a candidate has fifteen years of experience, he or she can actually be certified without any college education.

As AICPA president, I always tried to make the case that the value of on-the-job training had increased in importance as the types of employment open to the new CPA had expanded. College training gave students a strong foundation for learning, as well as an in-depth knowledge of fundamental accounting principles. But no undergraduate curriculum could possibly prepare a prospective CPA for all the many employment possibilities open to him or her—everything from audits, to business evaluation, to tax preparation, to financial planning, to mergers and acquisition. Specialization had to occur in a formal educational setting, after which practical experience becomes crucial.

It may be crucial, but is also inevitable, which is why many of us at the AICPA never felt the experience requirement was particularly essential. Besides, there is no effective way to monitor the nature of experience one receives or the degree of supervision involved. Nevertheless, because many others, particularly those active in the large state societies, believed that some degree of experience should be demonstrated, we accepted the one-year requirement embedded in the UAA.

Chapter 6

Financial Reporting

Let's assume you have a few thousand dollars hidden under your mattress. You know there must be a safer place to put it, someplace where it would appreciate in value over time. It certainly isn't going to do that sitting where it is. You might even sleep better without that lump in your bedsprings.

But where should you invest? The stock market has been steadily rising, and you have a few favorite companies you've been tracking. So you order some annual reports and other financial information that the SEC requires all public companies to disclose. Then you begin poring over the data: income statements, cash flows, balance sheets, financial ratios, and the like. Lo and behold, you find that most of the same information seems to be available in every report you've ordered, making it fairly simple to compare the past performance of each company.

Or is it? As we shall see, notwithstanding significant improvements in financial reporting that have occurred during the past two decades, financial comparisons among companies can still sometimes be difficult. The history of the standard-setting process, however, is one marked by a continuing effort at eliminating these differences.

By way of background, we already have discussed how the securities acts of 1933 and 1934 established the SEC and gave it the specific authority to establish rules governing financial reports of public companies. The SEC was charged by Congress to ensure full and fair disclosure of the financial information investors needed to make an independent evaluation of a public company's financial health. But rather than take on the job itself, beginning in 1938 the SEC delegated the responsibility for establishing financial accounting and reporting standards to the accounting profession. Between 1938 and 1959, the AICPA's Committee on Accounting Procedure (CAP) issued fifty-one accounting research bulletins that, together with textbooks and other accounting literature, formed what became known as generally accepted accounting principles (GAAP).

As the needs of both companies and investors became increasingly complex, it gradually became clear that the CAP no longer had the resources to thoroughly investigate the various new standards that were needed. So in 1959, it was replaced by another part-time body, the Accounting Principles Board (APB), which was supported by a small but full-time research staff. In its fourteen years of existence, the APB issued thirty-one new standards.

Then by the late 1960s, there was again a concern within the business community that the APB had outlived its usefulness. Probably the most frequent criticism was that since the APB was made up largely of independent auditors, it was more beholden to the interests of clients than to the public. A more independent body was needed. Investors, creditors, and preparers needed to become part of the process, since participation of these groups was vital for the ultimate acceptance of proposed standards. There was also a growing recognition that in an increasingly complex business environment, a part-time committee simply did not have the resources to develop high-quality standards in a timely manner.

FINANCIAL ACCOUNTING STANDARDS BOARD

In 1971, the AICPA appointed two committees composed of accountants and other businesspeople to study different aspects of financial accounting standards. The first was the Trueblood committee, named after its chairman, Robert Trueblood, a former president of the AICPA who at the time headed the accounting firm Touche Ross & Co. It was asked to investigate who was using financial statements and for what purposes.

At about the same time, Francis M. Wheat, a former SEC commissioner and then a partner at the Los Angeles law firm of Gibson, Dunn and Crutcher, was asked to chair the AICPA's Committee on Establishing Accounting Standards. The Wheat committee, as it became known, was charged with coming up with recommendations for improving the *process* of establishing standards.

The Wheat committee concluded that the responsibility for accounting standards should stay within the private sector, but should be removed from the AICPA and vested in a full-time, salaried, independent body. As a direct result of recommendations by the Wheat and Trueblood committees, in 1972 the accounting standards-setting process was totally revamped. The

AICPA and other sponsoring organizations established an organization composed of a full-time FASB and technical support staff, both operating under the auspices of the Financial Accounting Foundation (FAF).

To assure the appropriate interaction with the business and financial communities, a Financial Accounting Standards Advisory Council (FASAC) also was established. The Council meets with FASB members on at least a quarterly basis.

Like the POB (see Chapter Four), the method of appointing the members of the FASB was an attempt to create an infrastructure that would ensure the standard-setting body of its independence. The FAF consists of sixteen trustees, twelve of whom are elected by representatives of FAF's sponsoring organizations—the AICPA; the American Accounting Association; the Financial Executive Institute; the Securities Industry Association; the National Association of State Auditors, Controllers and Treasurers; the Institute of Management Accountants; and the Government Finance Officers Association. The other four at-large members are appointed by the FAF itself.

The FAF, in turn, appoints the members of the FASB and its advisory council. It is also responsible for funding the FASB. About two-thirds of the FAF's income comes from selling publications and collecting licensing fees. The other third comes from contributions—half from public accounting firms and half from other organizations.

Since its inception, special interests have tried to influence the FASB's decision-making authority, and sometimes to undermine its authority. One strategy over the years has been to appoint trustees to the FAF sympathetic to a particular group's position on various issues. As a result, there have been a number of changes in the FAF's makeup over the years.

In October 1985, for example, after pressure for more than a year from various business groups to give businesspeople more representation in the standard-setting process, the FAF appointed a second representative to the FASB from business and industry—C. Arthur Northrop, a former treasurer of IBM. Ten years later, business was still clamoring for a greater influence. In early 1996, the FEI, the leading organization of corporate financial executives, was still writing to the FAF complaining that the FASB board was too big, moved too slowly, and too often reflected an antibusiness bias.

While business has always clamored for more representation, so have other groups. A 1987 GAO report, for example, criticized the lack of user involvement in the standard setting process. The SEC, too, was concerned that corporations and accounting firms were overly represented on the FAF board of trustees. In fact, SEC Chairman Arthur Levitt spent the last few years of my AICPA tenure lobbying the FASB to be more responsive to the needs of the investing public. Initially, the foundation rejected Levitt's argument that additional public representatives should be placed on the foundation's board. In 1996, former board chairman Michael Cook, managing partner of Deloitte & Touche and then FAF chairman, was still advising Levitt that accepting his proposal would mean "excluding from consideration many dedicated individuals who have the knowledge, experience, and perspective needed to best serve the public interest." Later that same year, the FAF finally agreed to replace two business members with two public members.

I tended to agree during these years with the view that the best way to retain the integrity and independence of accounting standard setting was to have a majority of foundation trustees be truly independent. It's certainly not a bad thing that today, one-quarter of the sixteen members of the FAF's board of trustees are public members.

FASB CAUGHT IN MIDDLE AND UNDER ATTACK

In addition to the makeup of the FAF, other battles were also being fought. The FASB was established within the private sector as a way of developing fair and relevant accounting and reporting standards, free of political and special-interest group influence. It didn't exactly work out that way, however.

The FASB was continually besieged with recommendations to amend or add accounting standards. But it was also constantly under pressure not to overload preparers, users, and auditors with too many standards. (Many in the business community and the public accounting profession itself believed that there was just so much change in financial reporting that can be absorbed within any finite period of time.)

On the one hand, the SEC expected the FASB to be the protector of the small investor who didn't have the "clout" to compel public companies to provide information. After all, about half the securities traded in the capital markets are owned by individuals. The other half are held by institutional investors, who ostensibly have their own resources to investigate financial performance and are therefore not as dependent on the information that companies are required to disclose. Users (particularly individual investors) of the financial information invariably welcome new information that either voluntarily or by regulation is added to financial reports. Why wouldn't they? The more data provided to them, the better opportunity they have for determining the financial worthiness and future prospects of the companies in which they are considering investing.

Preparers, on the other hand, want to limit the kinds of financial information they are obligated to provide. Businesses are naturally interested in putting their best foot forward, so they often lobby for accounting standards that will reflect their companies' best attributes. Many also feel that every disclosure has some kind of built-in risk.

"Public companies are always concerned about divulging proprietary information," explains Gary Previts, a professor in the Department of Accountancy at the Case Western Reserve University's Weatherhead School of Management, the author of several books about the accounting profession, and a former AICPA board member. "Take American Greetings, located in my own city of Cleveland. As a public company, it is required to disclose all sorts of financial information. But its chief competitor, Hallmark, a private company, is not. American Greetings undoubtedly believes that disclosure rules put it at a competitive disadvantage, even though their argument may be less compelling than it once was because companies today know so much more than they ever did about their competitors."

Within this contentious environment, it has always been difficult for the FASB to determine which issues to investigate, much less which ones to ultimately promulgate as new standards. During my AICPA tenure, I repeatedly suggested that the FASB attempt to set an agenda for itself based on a longer-term assessment of user needs, rather than the more traditional approach of selecting items for study from among a wide array of possible subjects. In other words, I favored a more proactive, strategic process rather than one that simply reacted to the onslaught of suggestions thrown at it. A modus operandi could be created patterned after the AICPA's own strategic planning process.

From the FASB's point of view, this was easier said than done. Complicating matters was that the FASB did not actually have any direct legal authority to set standards. Rather, it relied on the support and concurrence of the SEC, which itself had to answer to both the legislative and executive branches of government.

Given these divided interests, it was no wonder that during these years the FASB's independence was increasingly threatened by special-interest groups that believed it could circumvent its authority. Usually that meant appealing to Congress whenever the FASB took a position not to the special-interest group's liking.

During the mid-1980s, for example, banks and insurance companies spent several years petitioning the SEC, often through their powerful congressional friends, for exemption from an FASB provision requiring more accurate reporting of deferred tax benefits. In this case, the FASB managed to withstand congressional pressure. But the FASB didn't fare as well during its next major battle.

When in 1983 the FASB proposed requiring executive stock options to be recognized as an expense, certain members of the business community, led by the burgeoning ranks of technology companies, protested that this would unfairly drive down their bottom lines. When the FASB refused to back down, intensified corporate lobbying led to a Senate resolution asking the board to find a compromise solution. Only after the SEC and the AICPA's Accounting Standards Executive Committee (AcSEC) joined the call for the FASB to modify its proposal did the FASB backpedal. In the end, it agreed to accept the disclosure of the estimated value of stock compensation.

Sometimes the FASB did prevail in its battles with special-interest groups. It was, for example, more successful in requiring more detailed reporting of postretirement employee benefits. Leading this effort was Dennis Beresford, who in 1987 succeeded Donald Kirk as chairman of the FASB. Denny was the former national director of accounting for Ernst & Young, chairman of the AcSEC from 1982 to 1984, and one of the AICPA's two representatives on the International Accounting Standards committee. These credentials gave credibility to his support of a program which caused many companies to reexamine their approach to reporting benefits and to highlight the present and future costs associated with their benefit plans. Many companies were persuaded to ensure that their plans were on solid footing in both appearance and fact.

DERIVATIVES

During the mid-1990s, the FASB again came under fire when, after years of study, it proposed a requirement that companies recognize derivatives on their balance sheets at market value. Derivative contracts are used by large businesses to hedge against fluctuations in interest rates, currencies, or commodities. A bank, for example, might use interest-rate derivatives to hedge the value of its loans. If rates move up, the value of the bank's loans is reduced, but this is offset by the increased value of the derivatives.

Derivatives are complex financial instruments that can be an effective risk management tool. But they also can expose an institution or company that trades in them to potential financial disaster. Large losses can accumulate quickly.

The most spectacular cases of derivative losses occurred in 1994. First Proctor & Gamble was forced to take a $157 million pretax loss from two risky types of derivatives. Then Orange County, California, declared bankruptcy, an event precipitated by risky derivative investments.

For years, many of us knew something had to be done about derivatives. In both the Proctor & Gamble and Orange County cases, the typical investor had no reasonable opportunity to be forewarned. The investments had not been disclosed on the financial reports because the companies were under no obligation to do so.

Supported by an understanding between the SEC and the accounting profession, the FASB had the responsibility to establish ground rules to adequately inform the investing public about all financial matters critical to a public company's continued viability. That certainly included any financial risk it took on. The FASB, therefore, rightly proposed that the underlying value of the derivatives themselves be reported on financial statements.

On one side of the controversy were banking institutions. Supported by Federal Reserve Chairman Alan Greenspan, they wanted at least to segregate derivative information on a different page, away from the profit-and-loss statement. They argued that reporting the value of derivatives would make balance sheets more volatile. But financial reporting doesn't create volatility. It only unmasks it. Most of us didn't see why it was unreasonable to make the investing public aware when a public company had placed itself in a precarious financial situation.

Congress had the authority to force the FASB to reverse itself on the derivatives issue—or on any issue, for that matter. Indeed, it could have further crippled the FASB by passing legislation, introduced in the U.S. House of Representatives, that would have required the SEC's approval of any FASB proposal after soliciting public comment. This would have completely gutted the FASB's ability to issue objective standards.

The derivatives debate illustrated the continuing threat over the independence of the FASB. If special-interest groups successfully appealed to Congress every time the FASB issued a new standard not to their liking, standard setting would, at least on a de facto basis, move from the accounting profession to the federal government. This idea troubled many of us.

The last time the federal government became involved in setting accounting standards was during the S&L crisis of the 1970s and 1980s. In a spectacularly unsuccessful effort to prop up failing banking institutions, it led to one of the biggest taxpayer bailouts in U.S. history. Giving the federal government free rein would also likely make accounting standards look more like the 10,000-page labyrinth we call the IRS code, rife with social agenda and special-interest priorities, than the measured body of rules and regulations that have been developed by the profession since first being given the authority to set accounting standards sixty-five years ago.

U.S. capital markets are the envy of the world, to a large extent due to reliable GAAP. Adherence to these standards translates into financial statements that can be relied on by the millions of small individual investors who form the backbone of our economy. It is the FASB's responsibility to protect them by not allowing the interest of any particular group to take precedence over those of the investing public.

JENKINS COMMITTEE

By the mid-1980s, most of us at the Institute recognized that accounting standard setting was not keeping up with the rapid changes taking place within the national and international business environments. As detailed in Chapter One, within the flurry of day-to-day activities the strategic planning process helped the Institute become more future-oriented. That meant keeping our focus continually trained on the most important issues identified by that process. One of the issues we always felt deserved further examination was financial reporting.

As the entire global business structure became more complex, there was little doubt that self-regulatory oversight needed to be strengthened and that accounting and auditing standards were in some cases no longer adequately providing the public with the information necessary to assess a company's financial condition. After all, the sheer number of people who relied on financial information had by the 1980s increased exponentially. From 1965 to 1990, stock ownership among Americans doubled, from 10.4 percent to 21.1 percent of the total adult population. It then took only seven years for it to double again, to about 40 percent. At the same time, the technological revolution substantially increased the amount of information available to the average investor.

The Institute's Future Issues Committee had on more than one occasion concluded that the information found in financial statements was becoming less significant as other kinds of information were increasing in importance. In 1988, our Future Issues Committee report entitled *The Changing Significance of Financial Statements—The Relative Disparity Between Content and Needs* declared that financial reports were losing their significance because they were not future oriented and therefore did not provide value-based information.

Compounding the pressure to take action, negative media attention was centering around charges that the accounting profession had not provided adequate early warning of bankruptcies and other fiscal problems within public corporations and financial institutions. Many of these charges focused on historical financial statements that did not necessarily provide the types of information users were seeking.

It wasn't that the FASB didn't have help in trying to address the most important financial reporting issues during these years. Both the FASB's Emerging Issues Task Force (ETIF) and the AICPA's AcSEC were there to supplement the FASB's ability to keep up with changing developments. (The AcSEC's accounting pronouncements tend to be narrow in scope and related to specific industries, and must be cleared by the FASB before going into effect.)

Created in 1984, the ETIF was really an acknowledgment that the FASB alone was not equipped to keep up with a business environment that was changing so rapidly. With the cooperation of the SEC (the SEC's chief accountant is an ex officio participant), the ETIF attempts to resolve issues quickly by reaching a consensus on how public companies should report new kinds of transactions. The SEC then immediately requires publicly registered companies to follow these new rules.

Nevertheless, the AICPA leadership believed another group was needed to take a strategic look at where financial reporting should be heading, and what could be done to provide the types of information critics were calling for. Probably the biggest proponent of setting up a special committee on financial reporting was Tom Rimerman, who during his term as chairman between 1990 and 1991 made financial reporting reform his top priority. Tom also brought to the Institute extensive firsthand knowledge of litigation support, which served to strengthen and support that specialty at the national level.

In the spring of 1990, Tom published an article in the *Journal of Accountancy* entitled "The Changing Significance of Financial Statements," in which he called for a new blue ribbon commission to study the relevance of financial reporting. In the fall of that same year, a symposium at the Wharton School also indicated the need for change.

Tom was instrumental in persuading Ed Jenkins to be chairman of this new committee. We couldn't have found anyone more qualified. Ed had been with Arthur Andersen since 1958, and as managing partner of the firm's accounting and audit practice he had worldwide responsibility for the largest segment of the firm's professional practice. Between 1984 and 1991, he had been a member of the ETIF, and then of its advisory council for four years after that. (In July 1997, Ed would become chairman of the FASB.) Other members of the Jenkins committee, as it became known, included representatives from the six largest accounting firms, five representatives from medium and smaller firms, two CPAs from industry, and an academic.

At our September 1993 Council meeting, Ed laid out the rationale for the special committee's efforts:

- The world is rapidly changing and external reporting is not changing to keep pace.
- Traditional external reporting was not designed to predict the future, yet investors and creditors expect information that helps them do just that.
- External reporting focuses on the past, yet the pace of change is accelerating so rapidly that the past is less likely to be a good predictor of the future.
- Because of dramatic leaps in technology, there are many emerging industries with a limited history. For them, reporting only on the past is not very helpful.
- Financial statements remain an important part of the users' analysis, but only a part. Many companies now focus on an ever-widening array of nonfinancial measures of performance to assess the cost, quality, and innovation of their products and services.
- Many companies divide their activities into business segments, so focusing only on the consolidated total often fails to provide a clear picture of the very different activities within the same company.

Under Ed's able leadership, the Jenkins committee managed to combine a vision of what would be needed in the future with support for the traditional framework of U.S. financial reporting. In 1993, it issued a comprehensive, 200-page final report entitled *Improving Business Report-*

ing—A Customer Focus. In essence, it concluded that while there was a lot right with financial reporting, changes were desperately needed.

The report's detailed recommendations could be synthesized into two broad categories, both intended to meet the profoundly changing needs of users of financial information. For one thing, business reporting needed to provide more information with a forward-looking perspective, including management's plans, opportunities, risks, and uncertainties. It also had to give greater emphasis to the factors that create long-term value, including nonfinancial measures indicating how key business processes were performing.

The report's second broad observation was that participants in the business reporting process needed to do a better job of anticipating change. They could do this by being more farsighted—by adopting a vision of the future business environment and of the kind of information investors would need to make intelligent choices. Toward that end, the report provided a detailed example of a new business reporting model.

In presenting the committee's final report at our October 1994 Council meeting, Ed responded to concerns from Council members about companies being required to report too much new information. He explained that companies should report only the information they have, and should be under no obligation to garner information they don't need to manage the business. They also should not be required to expand the reporting of forward-looking information until there were more effective deterrents to unwarranted litigation.

Ed also explained that management should not be required to report information that would significantly harm a company's competitive position. The goal was not to require forecasted financial statements, but to persuade management to provide information that helped readers forecast for themselves.

Another insightful analysis of the Jenkins committee's conclusions came from one of its committee members, Bob Elliott, a senior partner at KPMG Peat Marwick and later an AICPA board chairman. "Our accounting paradigm represents a well-developed structure for the industrial era," Bob noted at the AICPA-sponsored Wharton School symposium held at the University of Pennsylvania in October 1990. "But it is completely inadequate for the information era."

Bob pointed out that management's job was no longer focused primarily on a fixed basket of assets bequeathed to it by prior management, such as raw materials, finished goods, and plant and equipment. Now it had to focus on information assets. Training, research and development, market studies, planning, design, advertising, internal communications had all become an increasingly large proportion of the value delivered to customers. Accountants were nevertheless continuing to supply both management and investors with the same industrial-era financial statements.

"Accounting has been called the language of business," Bob concluded. "But when you read about modern management and business in the financial press or in periodicals or books, you find another language has taken its place."

STANDARDS OVERLOAD

As the U.S. tax code became almost unbelievably complicated, and as financial instruments and reporting requirements became equally complex, businesses were forced to comply with a seemingly never-ending number of regulations and standards. The resulting paperwork associated with preparing even the simplest of financial statements became daunting, particularly for small businesses without the resources of, say, Fortune 500 companies which had their own internal accountants. Add to this other challenges that confronted businesses, including keeping up with technological changes and constantly evolving labor and pension laws, and many companies cried out for relief.

Standards overload has always been of particular concern to small businesses, since they are least able to bear the substantial costs associated with complying with increasingly complex GAAP. At various times, the accounting profession has tried to take steps to alleviate this burden. In fact, two separate reports, one released just as I was beginning my tenure as AICPA president, and another just as I was retiring, made a variety of recommendations that, if enacted, would have lessened the burden on small and medium-size accounting firms. Unfortunately, nothing much concrete occurred during the years between these two studies.

The problem has no simple solution. It pits the commonsense idea that small, private companies do not necessarily require the same reporting requirements as the largest multibillion dollar conglomerates, against the notion that the public has the right to be confident that *all* companies, regardless of size or ownership, have complied with GAAP.

This dilemma is by no means a new one. As far back as 1929, a special committee of the AICPA's predecessor organization, the American Institute of Accountants, addressed the issue in a pamphlet entitled *Approved Meth-*

ods for the Preparation of Balance-Sheet Statements. It noted the criticism, on the one hand, by some accountants that accounting rules were "more comprehensive than their conception of the so-called balance sheet audit," and on the other hand by other accountants that "the procedure would not bring out all the desired information."

In late 1980, the AICPA's Special Committee on Small and Medium-Sized firms (dubbed the Derieux committee after its chairman, Samuel A. Derieux, a former chairman of the AICPA's board of directors) concluded that the non-Big Eight firms, as well as their small and medium-sized clients, were burdened with an excessive number of accounting rules and regulations. A survey commissioned by the Derieux committee had found that regardless of firm size, almost 90 percent of all CPAs believed there was an accounting standards overload problem, although most believed it was more acute for nonpublic companies than for public companies. The committee also conducted a number of member forums in which specific instances of standards overload came to light. Members pointed out, for example, that many professional standards were designed for the public securities market and were irrelevant, or at least not cost-effective, for smaller companies which did not rely extensively on outside unsecured credit.

Members of small and medium-sized firms also pointed out the unfairness of discriminatory clauses in some loan agreements, which required that companies be audited by "a Big Eight" or "nationally recognized" firm. The committee suggested that the Institute go on record as opposing this type of discriminatory clause so that companies could make up their own minds about the size or type of firm it wanted to conduct the audit.

The committee also found that many smaller firms were fearful that the peer review process would require them to have the same quality control policies and procedures as the largest international firms. We tried to spread the word to them that the Division for Firms had already announced that peer review of firms with fewer than 20 professionals could be engagement oriented, rather than procedure oriented, and would be based on the policies and procedures appropriate for that firm. "The smaller firms don't have to adopt someone else's policies and procedures," Sam Derieux told us at the October 1980 Council meeting. "And they can have very good, sound policies and procedures that don't have to look the same as those found in the very large firms."

The Derieux committee recommended that the AICPA appoint a special committee to study how small businesses might be relieved from accounting standards that, given their size, were not cost effective. Such a committee was convened, and in December 1981 the Special Committee on Standards Overload issued its findings.

The special committee observed that smaller businesses did not have much discretion in applying standards. That's because they typically hired outside auditors, who were professionally obligated to adhere to all GAAP. The committee recommended reconsideration by the FASB of several accounting standards that it identified as particularly burdensome, and perhaps unnecessary, for small businesses. These included standards concerning leases, taxes, business combinations, and capitalization of interest costs.

Fifteen years later, however, substantive changes have not been made. It was a difficult issue to mediate. On the one hand, the problem for many small companies and accounting firms has only worsened during the intervening years. Tax laws and pension and labor reporting requirements, and therefore accounting standards, had only grown more complex. On the other hand, the expectation by the public and the government that financial statements fully divulge all pertinent information about a company's financial performance and prospects had also increased.

Still, the Institute was sympathetic to repeated complaints on the part of practitioners serving small and medium-sized firms. In January 1995, the PCPS established a special task force on standards overload. Its initial, tentative conclusions included some rather substantive recommendations for relaxing certain standards for small, privately held companies, although in the end the report's recommendations, issued in August 1996, were limited to helping small and medium-sized businesses comply with existing standards. This was a far cry from substantive relief, although the report did point to ways the AICPA could be helpful. The recommendations suggested, for example, that:

- Articles on each new pronouncement on standards should be published in the *Journal of Accountancy* in a succinct and understandable format relevant to small entities.
- Standards should be written in unambiguous and easy to understand language, and specific examples should be provided whenever appropriate.
- The hours and staffing of the AICPA's technical information hotline should be expanded.
- Relevant professional literature should be provided to members in a single CD-ROM format.

- Members should be provided the option of ordering a minimum library package when paying their annual dues.
- Access to professional literature should be made available to CPAs for a fee through the AICPA's fax hotline or via computer.

There were other suggestions during these years on how the profession might help reduce the burden of standards overload. Ray Groves, who helped build Ernst & Whinney (later Ernst & Young) into a global powerhouse, and who was chairman of the AICPA from 1985 to 1986, was always particularly vocal in this area. On his retirement in 1994, he made two particularly innovative suggestions, neither of which, unfortunately, was enacted. He recommended that every standards-setting agency, including the FASB, AICPA, and SEC, submit to a review every five years by outside experts charged with paring away deadwood rules. He also suggested limiting the annual report to a mere summary of financial results and the auditor's opinions, and placing the much more detailed information into a comprehensive supplementary report

Neither of these suggestions was implemented, and I frankly don't hold out much hope that any substantive relief from standards overload will be given any time soon to smaller companies. While it may be true that the number and intricacy of accounting standards are unfairly burdensome to small companies, companies that are publicly financed must adhere to SEC reporting requirements, as well as FASB and ASB standards, regardless of their size. There is more of a likelihood, however, that privately held companies not under the SEC's authority will eventually be relieved of some of the most burdensome standards.

A NEW PARADIGM

There is no doubt that financial reporting needs a new paradigm. Probably the greatest problem with the existing model is that it is based on an industrial economy in which companies were primarily involved with manufacturing and merchandising. In that environment, material, labor, and overhead were the primary ingredients. But in today's postindustrial service, knowledge-based economy, many valuable assets, including brand names, patents, copyrights, and internally generated research and development are not required to be reported. These costs are treated as expenses when they are incurred, rather than as assets to be written off as revenues are earned. The value of these intangibles, therefore, are largely ignored in financial reporting.

Current financial statements also pretend that inflation doesn't exist. They limit themselves to a mixed bag of historical cost and current value measurements. Although in recent years there appears to be a move in the direction of current value, the existing accounting model is still a hodge-podge of conflicting information. Some items, such as fixed assets, are recorded at their depreciated acquisition cost. Others, like marketable securities, are shown at market value. Since prices change over time, items carried at historical cost bear no relation to current value, the current cost to replace them, or even historical costs adjusted for changes in the purchasing power of money. Coupled with a variety of alternative accounting principles, such as FIFO (first in, first out) and LILO (last in, last out) for inventory, or accelerated and straight-line depreciation for fixed assets, it's no wonder that only the most sophisticated users can understand and interpret what in the world is going on.

In today's information age, companies also should be providing real time data to both investors and potential investors, much the same way theatrical film box office revenues are reported. "No one would conceive of driving a car with a television screen in place of a windshield, providing hourly snapshots of where he or she was two hours ago," is the imaginative way Bob Elliott describes it.

One of the central recommendations of the Jenkins committee was for more future-oriented, or forward-looking, information that would complement the historical data that have always dominated financial statements. Before passage of the Private Securities Litigation Reform Act of 1995, however, few public companies were willing to risk predicting future performance, lest they open themselves up to shareholder lawsuits if they failed to meet their stated goals. "Managements see disclosure of forward-looking information, even though helpful to users, as providing ammunition for future groundless lawsuits," the Jenkins report observed. "Many users also are concerned that unwarranted litigation is discouraging companies from disclosing useful information."

Fortunately, the litigation reform act now protects the preparers of financial reports by giving them "safe harbor" for the future-oriented information they include in financial statements. So at least this one excuse for inaction has largely been dissolved.

The Jenkins report was by no means the only study during these years that acknowledged the need for more relevant reporting standards. In 1991, for example, the FAF formed its own oversight committee to evaluate the operations of the FASB. The FASB itself in 1995 began a major research project to develop ideas for a new business reporting mode along the lines recommended by the Jenkins committee. One of the FASB's most interesting ongoing projects is its attempt to broaden compliance with voluntary standards.

AIMR REPORT

Another important study was the 1993 report, *Financial Reporting in the 1990s and Beyond*, published by the Association for Investment Management and Research (AIMR). The AIMR is a membership organization of more than 25,000 investment management professionals, including securities analysts, portfolio managers, strategists, and consultants. The purpose of the report was to disseminate AIMR viewpoints on matters relating to financial reporting and disclosure. Like Jenkins, and for that matter like virtually every other recommendation that found its way to the FASB during these years, the AIMR Report focused on three events that were rapidly making current reporting standards obsolete: globalization, computing power, and the move from a manufacturing economy to a service economy.

But most of all, the AIMR Report called on the FASB to be more aggressive in making financial statements more meaningful to the typical user. "We disagree with those who say its standards are too theoretical, that the cost of implementing them is too great, or that the FASB is inimical to the interests of financial statement preparers," it concluded. "Rather than following due process too little, we believe the FASB follows it too much."

STATE AND LOCAL GOVERNMENTAL UNITS

In the wake of New York City's financial crisis, during the mid-1980s legislation was introduced in Congress to establish a governmental accounting standards board as a new federal agency. The accounting profession and state and local government officials strongly opposed federal control of the setting of such standards, and cooperated in having the FAF instead establish a Governmental Accounting Standards Board (GASB) within the private sector.

The FASB wasn't too pleased with this arrangement either, although surely it was better than having a new federal agency do the job. Yet the FASB believed it was perfectly able to set the standards itself for all nonbusiness units, whether public or private.

On the other hand, throughout the 1980s state and local governments argued that government operations required special treatment, and that a separate body should set rules for them. The AICPA and FAF supported the establishment of the GASB, as long as certain criteria were met. First, the board had to be truly independent, not controlled by any arm of the government. Second, it had to offer reasonable prospects for improved financial reporting by state and local governments. Third, there had to be reasonable assurance that adequate financing would be made available on a continuing basis. And fourth, a mechanism had to exist for resolution of potential conflicts of jurisdiction between the GASB and the FASB.

In 1984, over the FASB's opposition, we believed these criteria had been met as the FAF established the GASB to set standards of financial accounting and reporting for state and local government units. As with the FASB, the FAF was made responsible for selecting the GASB's members, funding their activities, and exercising general oversight.

Although GASB pronouncements may have had a lower public profile than many recent FASB proposals, they too were rarely finalized during these years without some controversy. Toward the end of 1986, for example, public employee retirement systems objected to a new GASB rule requiring states, cities, and towns to disclose more pension data. Under the old method, there were a dozen or so different funding and reporting methods for pension funds. That made it difficult for financial analysts to compare government securities and determine which ones to purchase. Ironically, the retirement systems were concerned that the rule would make their pension funding look better than it really was. That, they argued, could lead legislators, employee groups, and unions to seek added employment benefits that the states couldn't afford. At the AICPA, we believed this was a rather disingenuous argument.

On at least one occasion, the FASB and the GASB also clashed over jurisdiction. In the summer of 1987, the FASB issued a rule requiring all colleges to depreciate buildings and equipment. A few weeks later, the GASB exempted state-owned educational institutions.

Depreciation is an accounting technique used to write off the cost of an asset over its useful life. Since it reduces the book value of assets, it can lower overall profit. Private and public colleges compete for financing by issuing tax-exempt bonds, so private schools complained that the new rule would put them at an unfair disadvantage when trying to raise money.

The issue simmered for more than two years until November 1989, when the FAF voted to require government-owned hospitals, utilities, colleges, and universities to follow the rulings of the FASB. Almost immediately ten state and local government groups threatened that if the ruling was not reversed, their state legislatures would promulgate accounting rules that would supersede those made by the FASB. The fact that the ten government groups together supplied more than $1.1 million of the GASB's then $1.8 million annual budget only strengthened their bargaining position. Ultimately, the FAF backed down and agreed to adhere to its original bylaws by which the GASB could overrule the FASB when it came to accounting standards for government-controlled entities.

FINANCIAL ACCOUNTABILITY OF THE FEDERAL GOVERNMENT

Another focus of the Institute during these years was an effort to promote legislation that would make the federal government as accountable for its income and expenditures as was demanded of any public company in the United States. Two people in particular were instrumental in promoting a group of laws that did just this. Joe Moraglio was the AICPA's vice president of federal government affairs during most of these years, and Charles A. Bowsher, a CPA, served a fifteen-year term as U.S. comptroller general that almost exactly coincided with the period of time discussed in this book. In fact, Chuck Bowsher has to be considered the single most important figure in the effort to reinvent government finance, since under his leadership the GAO became involved in some of the most significant accountability issues in recent U.S. history, including passage of the Single Audit Act and the Chief Financial Officers Act, and in helping to form the Federal Accounting Standards Advisory Board, which would eventually seek authoritative status as a GAAP determining entity.

Prior to enactment of the Single Audit Act in 1984, every federal grant program had its own audit requirement. If a U.S. city, for example, received 300 grants from various agencies of the federal government, it was apt to have not too many less than 300 different auditors. There was scant uniformity among the audits, and no centralization of the overall results. The Single Audit Act eliminated this confusion by legislating that each municipality must undergo one audit covering all federal monies it received.

The AICPA supported enactment of the Single Audit Act, but we were also sensitive to the concerns of smaller practitioners who feared they would be eliminated from the process, since many audits of municipalities now required a larger firm to conduct the single, larger audit. We were able to persuade Congress to include language in the bill that consideration should be given to subcontracting smaller firms to conduct some of the work.

In 1990, we were much more active in supporting the Chief Financial Officers Act. It required the top twenty-four agencies of the federal government to appoint a chief financial officer, who would be responsible for producing audited financial statements to account for all their agency's activities, spending, and revenue. The bill established a chief financial officer within the Office of Management and Budget (OMB), as well as financial officers in most government agencies. These individuals were made responsible for preparing financial statements using uniform accounting principles, which would receive an annual independent audit. No such financial oversight had previously existed in the federal government.

Much of the Chief Financial Officers Act stemmed from the work and recommendations of a special AICPA task force that held a colloquium on improving federal financial management in Washington in December 1989. The task force succeeded in garnering the support of the House Government Operations Committee and the Senate Governmental Affairs Committee, and the bill passed with strong bipartisan support.

The Institute and the bill's other proponents recognized the difficulty of treating the federal government like a normal business. Should, for example, Yellowstone National Park be listed as an asset worth what a developer would pay for it, or should it be evaluated using some other measure? How could an expenditure for an entitlement program, which had no asset value, be measured in the same way as an expenditure by the military for a tank, which could be considered a capitalized asset?

Despite these difficulties, for the first time in our country's history the Chief Financial Officers Act succeeded in making the federal government financially accountable. Nevertheless, after it became law, it almost immediately came under attack, most notably from House Appropriations Committee Chairman Jamie L. Whitten, a Democrat from Mississippi, who effectively tried to kill it by failing to appropriate the necessary funds for its implementation. In response to a Bush administration request of $105 million to fund 23 personnel positions called for in the act, Whitten wrote House Government Operations Chairman John Conyers, Jr. (D-Michigan) that it created an unnecessary layer of bureaucracy and was an intrusion by the executive branch into the affairs of Congress. "It is my judgment that the new financial officers would make our job harder, would result in less control of the Congress over its own operations and would, to a degree, further turn congressional decision making over to the executive branch of government," he wrote. Whitten further argued that budget officers currently in place in each department and agency were sufficient to keep track of financial matters.

The Institute could not have disagreed more strongly with Whitten's position. The legislation was an important first step toward financial responsibility on the part of the federal government. Indeed, for the first time since the birth of the Republic, accountability on the part of a designated financial officer had become a reality. In a statement drafted by our Washington office, we declared that "Thomas Jefferson called for federal financial oversight two hundred years ago. The Chief Financial Officers Act of 1990 is the first time Congress has responded to that call. We can't afford to back down on it now. With all due respect, we believe that Mr. Whitten has got it wrong. He is against a law which will, after 200 years, finally bring accountability to government." Fortunately, in the end the act was fully funded by Congress.

The concept of federal budgetary accountability was further strengthened in 1994, when Congress passed the Government Management Reform Act. It required the Office of Management and Budget to publish, beginning in 1998, an annual financial statement covering all agencies of the executive branch.

Taken in total, this group of federal legislation went a long way toward putting our federal government's fiscal house in order. Although the first group of audits exposed the government's sloppy accounting practices, that was part of their purpose—to promote accountability so that each year's financial statement would be steadily improved. That has proven to be the case.

RESISTANCE TO CHANGE

When a court decision leaves both sides dissatisfied, yet still not clamoring for an overthrow of the entire judicial system, it may indicate that the decision has been balanced and equitable. Using that analogy, the FASB must be doing a superb job.

On the one hand, it is the rare new pronouncement from the FASB that does not outrage one or more constituencies. On the other hand, none of the FASB's constituents are suggesting that investors would be better served if standard setting was taken over by the federal government.

Still, sometimes it seems that the FASB is continually under siege. Take your pick: The FASB is moving too slowly; it's moving too quickly. It's not doing enough to make financial reports more meaningful to their principal users; it's overburdening preparers with too many standards. These statements reflect the continuous tug of war that has deflected the FASB's power to implement the paradigm shift that many believe is so sorely needed.

Resistance to change comes from three sources: auditors, financial analysts, and the actual preparers of the financial information. One would think that auditors would support expanding the financial reporting model, since if more information was required to be confirmed by an audit, accounting firms would theoretically have more work to do and could raise their fees. But as a group, auditors are often satisfied with the status quo. Perhaps it is because forward-looking information is more susceptible to error than historical information, and that historical information is easier to quantify.

Financial analysts have the resources and sophistication to develop their own forecasts, so they have little incentive to support having companies compile forward-looking, more relevant information that would then be available to everyone. As for those who actually prepare the financial information, they want maximum flexibility in their ability to tell their story. They therefore don't seem particularly interested in an expanded reporting model, unless it is obvious beforehand that such disclosures would reduce the cost of capital.

So it is a continuous tug of war, with the force in the middle keeping the rope stationary, emerging as the winner so far. Compounding the problem is the well-intentioned but slow-moving process that the FASB uses to consider new standards. The FASB operates under rules of procedure that require an extensive due process open to public observations and participation. While only one opportunity for public comment is provided for in the rules, often the FASB goes well beyond that in an attempt to form a consensus. This is particularly true when the proposed standard is controversial, as many standards are. After first proposing standards for reporting derivatives, for example, the board gave its constituents five separate opportunities for comment, which helped string out the process for at least a decade. Its project on pensions and other postemployment benefits took even longer to conclude.

AN INTERNATIONAL SOLUTION

All these forces together make one pessimistic that any meaningful expansion of the financial reporting model, including those changes recommended by the special committee on financial reporting in 1993, will occur any time soon. Complicating the situation further is the increasing importance of the International Accounting Standards Committee (IASC), which has been working on a set of core accounting standards to which all its members would adhere. The IASC was formed in 1973 through an agreement made by the professional accountancy bodies of Australia, Canada, France, Germany, Ireland, Japan, Mexico, the Netherlands, the United Kingdom, the United States, and Ireland. Since 1983, IASC's members have included representatives from the eighty-nine countries that belong to the International Federation of Accountants.

No profession is more international in scope than accounting, but differences in how countries regulate accounting have resulted in barriers that prevent the profession from meeting the changing needs of an increasingly global economy. As business becomes increasingly global, and as cross-border mergers become more commonplace and international capital markets expand, it becomes more and more imperative that a single set of international accounting standards be established to permit the unencumbered flow of investment and commerce around the world. Likewise, the explosive growth of international capital and investment, as well as the volume and size of international securities offerings, has fueled a demand for a core set of high-quality international accounting standards that business enterprises from all countries can use when seeking access to the international capital markets.

Already, many countries use the IASC's existing rules as a basis for their own national accounting requirements. Emerging economies regularly use them as a benchmark when developing new rules. They are also used by many stock exchanges and regulatory authorities so that foreign companies can present financial statements in a consistent, meaningful way. But there are still plenty of gaps in the IASC's rules and areas for which it has not yet issued standards.

Given the growing market for international securities, over the years the FASB has sought to play a leading role in working with the IASC and other nations' accounting associations to discuss the need for worldwide standards. By 1995, however, a substantial amount of uncertainty still remained as to the ultimate outcome concerning which body would ultimately establish international standards. While the FASB continues to be the premier standard-setting body in the world, and while the quality of financial reporting in the United States is second to none, there is a natural resistance on the part of other nations to simply accept FASB standards.

Eventually, one set of global financial reporting standards will undoubtedly emerge to satisfy the needs of the global capital markets and business communities. My guess is that the marketplace will prevail, and that the IASC or some other successor body will emerge as the global standard setter, with national standard-setting bodies such as the FASB playing only a supporting role. Indeed, as I conclude the writing of this book, developments of this type are approaching consensus and conclusion.

Chapter 7

Changing Profession, Changing AICPA

Between 1980 and 1995, AICPA membership doubled, from 161,319 to 323,779. Of equal significance was a pronounced shift toward more AICPA members from business and industry. In 1980, members working in accounting firms outnumbered those in business by about five to three. But as can be seen from Table 7.1, by 1995 this dominance had ended, and precipitated a trend that became even more pronounced in subsequent years.

This demographic shift within our membership led to a subtle shift in the emphasis of the kinds of direct services we provided our members. More CPE courses, for example, were directed specifically to the interests and needs of CPAs in business and industry. Despite these internal changes, our external activities did not change significantly. Serving the needs of the CPAs in public practice and speaking out publicly on their behalf enhanced the status of the entire profession. After all, there has always been considerable movement back and forth by individual CPAs between public and corporate practice.

Stephen J. Young of Arthur Andersen in Chicago documented some of these changes in a 1995 article he wrote for the journal *Research in Accounting Regulation*. Entitled "The Changing Profile of the AICPA: Demographics of a Maturing Profession," Young identified four major demographic trends relating to AICPA membership:

- After many decades of rapid increases, membership growth was finally slowing. It had peaked at almost 12% per year during the post-World War II period, between 1945 and 1955. For the next three decades, annual increases had averaged between 6.5 and 8.5 percent, until

165

Table 7.1. AICPA Membership Changes 1980–1995

	1980	1995
Total membership	161,319	323,779
Public accounting firms	54.1%	40.7%
Business and industry	35.5%	41.7%
Education	2.9%	2.4%
Government	3.3%	4.4%
Retired and others	4.2%	10.8%
Public accounting firms	87,339	131,887
Firms with 1 member	23.8%	23.2%
First with 2–9 members	33.1%	36.5%
Firms with 10+ members	13%	20.4%
25 largest firms	30.1%	19.9%

falling to 5.5 percent between 1985 and 1990. From 1990 until 1994, it then dropped precipitously, to 2.2 percent.

- By 1994, the number of members in corporate practice was, for the first time, equal to the number in public practice.
- More and more women were becoming CPAs. In fact, in the youngest age category, women outnumbered men.
- There had been a "massive growth" in the number of accountants, compared with engineers, lawyers, and doctors. This was "consistent with the increased demand for information required in a modern economy."

According to Young, these trends suggested "a membership which is changing in fundamental ways, creating new challenges for the leadership of the Institute." To the profession, his most encouraging finding documented what had become increasingly evident to the AICPA leadership. Accountants had "become a more important part of our modern information-based society," the report concluded.

DOCUMENTING PERCEPTIONS

The reason we had become a more important part of the business landscape was that accountants were now offering a wide variety of business services, some of them far removed from the traditional audit or tax function. As these changes were being made, we believed it was important to begin to document how we were perceived by our various constituencies. Only then would we hope to attempt to change any perceptions we believed were inhibiting our efforts to branch out into new areas.

One of the first such efforts was made by the accounting firm Peat, Marwick, Mitchell & Co. In 1984, it hired the Opinion Research Corporation (ORC) to assess the overall reputation of independent auditors and to determine how well the profession's role and responsibilities were understood by those it served.

The ORC survey results were, at first glance, quite heartening. Eighty-seven percent of businesspeople, which included chief executive officers, security analysts, and regulatory officials, agreed that auditors had high ethical and moral standards. Ninety percent said that auditors exercised independent and objective judgments in performing audits of financial statements, and 72 percent thought that the profession was doing a "good" or "very good" job. Most significantly, CPAs were ranked higher in these characteristics than all other professional groups, including university professors, bankers, doctors, lawyers, stockbrokers, insurance agents, television newscasters, and U.S. senators and congressmen.

This was heady stuff, particularly since at the same time we were being given such high marks from businesspeople, we were simultaneously being assaulted in the media and in Congress for our culpability in a handful of high-profile business and bank failures that were occurring during these years (see Chapter Two).

The Institute recognized a need to build on the ORC study, and to investigate in further detail attitudes toward our profession. It was one thing to receive high marks for integrity and for competently performing our traditional tax and attest functions. But were these perceptions also limiting our ability to expand into other areas? After all, most of us agreed that such an expansion was crucial to the growth and continued vibrancy of our profession.

Consequently, in 1986 the AICPA commissioned Louis Harris and Associates to conduct a comprehensive survey of the principal publics that we served. These included both the American public and "leaders," defined as business executives, bankers, attorneys, legislators, security analysts, and members of the financial media. (See Chapter Four for a discussion on how the Lou Harris survey was used to support our decision to sign the FTC consent decree concerning commissions and contingent fees.)

No doubt the most heartening finding of this new study was its confirmation of the ORC's results (see Table 7.2). CPAs were indeed held in high regard compared with other professions.

Table 7.2. Ethical and Moral Practices of Key Occupations

Question: "How would you rate the ethical and moral practices of people in the following professions and occupational groups?"

	Leaders			Public		
	Positive	*Negative*	*Not sure*	*Positive*	*Negative*	*Not sure*
CPAs	90%	9%	2%	57%	21%	22%
College/university professors	81%	14%	4%	58%	25%	17%
Bankers	78%	20%	2%	59%	33%	8%
Doctors	78%	20%	2%	64%	33%	3%
Corporate executives	70%	27%	3%	33%	40%	27%
Editors of newspapers and magazines	52%	43%	4%	47%	42%	11%
U.S. senators and congressmen	49%	48%	3%	34%	58%	8%
Television newscasters	45%	51%	4%	60%	34%	6%
Stockbrokers	43%	51%	6%	33%	31%	36%
Personal financial planners	41%	48%	11%	42%	31%	27%
Insurance agents	43%	53%	4%	44%	48%	8%
Lawyers	43%	54%	2%	40%	50%	10%

Note: Annotated from "A Survey of Perceptions, Knowledge, and Attitudes Towards CPAs and The Accounting Profession," prepared for the AICPA by Louis Harris and Associates, Inc. 1986.

But the importance of the survey results was not that people held us in high esteem. We already knew that. Our paramount goal was to address the concern on the part of our rank and file members that CPAs were not taken seriously enough in our critical attempt to expand from our traditional attest and tax duties.

At first glance, our fears seemed misplaced. Large majorities of business leaders agreed that it was appropriate for an auditing firm to be engaged in assisting in computer hardware and software selection and design (80 percent) and general management consulting (79 percent). However, when compared with management consulting firms, respondents indicated significantly less confidence in the ability of CPAs to act as financial advisors to corporations. A majority (57 percent) also believed that serving as an overall business advisor affected the ability of an independent auditing firm to be objective when auditing a client's financial statements. Particularly disturbing to many of us, although not particularly surprising, was that while the survey results demonstrated extremely favorable attitudes toward CPAs, the one characteristic for which we were not given high marks was for our creativity (see Table 7.3).

Table 7.3. Rating of CPAs on Key Attributes

(Question: "In general, how would you rate CPAs on ...")

	Leaders			Public		
	Positive	Negative	Not sure	Positive	Negative	Not sure
Honesty	89%	9%	2%	61%	28%	11%
Competence	86%	12%	2%	65%	24%	11%
Reliability	85%	13%	2%	65%	25%	10%
Objectivity	79%	20%	2%	56%	29%	15%
Concern for the public interest	55%	43%	2%	52%	39%	9%
Creativity	39%	57%	3%	50%	34%	16%

Notes: Annotated from "A Survey of Perceptions, Knowledge, and Attitudes Towards CPAs and The Accounting Profession," prepared for the AICPA by Louis Harris and Associates, Inc. 1986.

Furthermore, the report's analysis observed:

> Put bluntly, CPAs do not generate a sense of being creative. In fact, it might well be that many CPAs have adopted a stance of studied avoidance of being creative. The reason might well lie in the notion that unless a CPA appears to be highly conservative and cautious, then others will get an impression of his or her being unreliable or unobjective. ...While having a reputation for being creative may not be as indispensable for CPAs as some other professions, ... nonetheless the utter lack of a solid reputation for creativity can be a serious handicap for CPAs. Most immediately, for a field which has now gone well beyond just the auditing function, to have to struggle to win credibility for creativeness can spell real trouble. This would be the case in management consulting, tax and financial advisory services, and a whole host of other areas.

The other important finding that confirmed for us that the profession needed to do a better job of marketing itself was the response to the question, "Do you think that the accounting and auditing profession needs to be better understood." An astonishing 77 percent of business leaders and 83 percent of the general public agreed that it did.

CHANGING PERCEPTIONS

At various times between 1980 and 1995, the AICPA initiated public relations initiatives that stressed the commitment on the part of CPAs to integrity and quality. We have already seen, for example, how in 1984 the

Division for CPA Firms embarked on a national public information program that included newspaper articles, advertisements, and speeches. They were all made in an effort to communicate the profession's high standards, highlight our dedication to quality practice and service, and describe how membership in the division assured the public that CPA firms would always have a strong commitment to quality.

The ORC and Lou Harris surveys subsequently demonstrated that whatever we had been doing to promote a reputation for integrity and competency was working. Yet at the same time, it was becoming increasingly clear that we needed a broader initiative to communicate the message that CPAs were also prepared to provide businesses with a number of other services outside our traditional tax and audit functions.

After the centennial year, which did generate some positive media attention, a very serious concern on the part of many members still remained. Nothing really had been done to change the perception on the part of the public or the business community that we were accountants and auditors, not creative businesspeople who could help in all aspects of business activity.

In 1986, the board of directors appointed a special committee on professional advertising with the charge "to consider the use of professional advertising as a means of educating the public about the CPA profession." Chaired by Ed Weinstein, the committee began by reviewing advertising campaigns by both small and large firms and by such associations as the Canadian Institute of Chartered Accountants, the National Education Association, and the American Hospital Association. It also reviewed the relevant literature, including a 1986 AICPA strategic planning survey that had found that 86 percent of all CPAs viewed a national advertising campaign as an important goal for the profession.

After being pitched by a number of ad agencies, the committee ultimately selected Bliss Barefoot & Associates, which had worked with the Institute on our communications efforts to promote the *Plan to Restructure Professional Standards* and on developing the Division for CPA Firms. At our May 1987 Council meeting, Bob Israeloff, who later became chairman of the AICPA board, explained that the actual ad campaign's overall objective was to "deliver the consistent message that AICPA members are the leaders of the accounting profession, emphasizing our professionalism and specifically the recent enactment of higher standards to which all members must adhere."

"To simply assert how good we are with ads shouting that we are better and more professional than ever is not the way to go," Bob explained. "This kind of approach invites our detractors to voice an opposing view. Instead, we need to focus on the broader issues that affect the entire business community, not just accounting. To do otherwise, would be, in effect, talking to ourselves. Placing issues that concern the accounting profession into the broader context of changing business conditions will enable the Institute to take the high ground."

Toward this end, the campaign, using a theme of "AICPA—the Measure of Excellence," pointed out AICPA achievements such as implementation of the Treadway commission recommendations, our restructured code of conduct, mandatory CPE, and peer review. It was aimed at three target audiences: business leaders and the financial community, the media, and, in an effort to stem any defections in our membership due to increased requirements, AICPA members. We budgeted $400,000 for fiscal year 1988 and $1 million for fiscal year 1989 so that the ads could be placed in the leading national business publications.

In March 1991, Geoffrey Pickard succeeded Bill Corbett as vice president of communications. Bill, who had spent sixteen years in charge of public and government relations at Avon Products, had joined the Institute in 1984. He boosted our public relations efforts considerably, and presided over the AICPA's centennial celebration, which included an expanded edition of the *Journal of Accountancy* and a variety of special events, including an expensive AICPA-sponsored entry in the Tournament of Roses Parade on New Year's Day 1987 in Pasadena. That event drew a number of responses and comments from the membership and the media—everything from good-natured ribbing to expressions of unbelieving dismay!

But Geoff came on board at a time when we were beginning to understand the need to make the profession more publicly visible. Geoff had held several senior level positions in advertising and communications at leading corporations such as General Electric and GTE. He brought with him an aggressive approach toward communicating our most important messages. By 1992, he had developed a plan to respond to an increasingly urgent plea on the part of many members that the profession increase its visibility. We all agreed that more media coverage was needed to expand our reputation as businesspeople, not just auditors and tax preparers. As a result, our public relations efforts were increased, with more proactive activities to help educate the media and more interview opportunities for senior members of the profession and of the Institute. Indeed, between 1992 and 1993, during

Jake Netterville's term as chairman, the accounting profession could be found in an increasing number of positive media stories.

Yet still, this didn't fully satisfy our membership. Even with the profession being portrayed in a more positive light in *some* stories, that's not what people tended to remember. The CPA's image continued to be largely driven by more negative stories that connected us with high-profile bank and business failures.

Toward the end of 1994, Geoff made a presentation to the board of directors suggesting a national advertising initiative. Nothing short of a major financial commitment to a comprehensive image enhancement campaign would serve our purposes, he told us. The board shared Geoff's enthusiasm and agreed to appoint a Special Committee on Advertising, charged with finding an advertising agency that could take our concepts and put them into a creative campaign. At the same time, we commissioned the national research firm, Audits & Surveys Worldwide, to create a benchmark of attitudes toward the profession so we could measure the effectiveness of any future advertising campaign.

The initial study by Audits & Surveys queried precisely the audience we needed to reach. Respondents were all key business decision makers in top management, mostly presidents and owners. They represented companies of all sizes and industries in all parts of the country. Eighty-one percent were directly involved in hiring outside consultants, including CPA firms.

Respondents were presented with various business services—everything from audits, to helping a company secure financing. They were then asked which type of professional—CPAs, bankers, lawyers, or stockbrokers—could best perform each function.

Not surprisingly, by wide margins CPAs were considered the best profession to prepare tax returns (95 percent), audit financial statements (90 percent), set up accounting systems (89 percent), develop the format for quarterly reports (87 percent), review and compile unaudited financial statements (85 percent), and suggest tax strategies (82 percent). There was even considerable support for CPAs being the best profession to help analyze operating results (66 percent), develop budgets and business forecasts (66 percent), and develop data processing and management information systems (50 percent).

But in other areas, there was considerably less of a belief that CPAs were the best professional group to consult on business problems (46 percent), assess benefit and compensation plans (44 percent), develop marketing and pricing strategies (32 percent), review personal investment strategies (30 percent), develop succession planning for family-owned businesses (28 percent), or help secure financing (14 percent).

At the same time we were establishing these benchmark data, we were interviewing ad agencies in an effort to develop a concept for an ad campaign that would effectively achieve our goals. As part of this process, we threw the question back to our members. In meetings throughout the country, aided by many state societies and often joined by professional facilitators, we challenged members to come up with a single definition of a CPA. Together we came to realize that this was no easy task. Not surprisingly, tax practitioners, university professors, auditors, consultants, and chief financial officers all had a different definition.

But even without closure on a common definition, there *was* a commonality in what we were hearing from our members. We clearly needed a creative national ad campaign that aggressively presented our most important message: that CPAs were uniquely qualified to provide an entire range of business services, not just the tax and audit functions for which we were best known. Our mission was to convey that CPAs were:

- More than tax and audit professionals, more than number crunchers
- Able to offer a broad range of skills and services to businesses and to individuals
- A vital ingredient in a company's success and bottom-line profitability
- Forward-thinking financial planners

After a careful review, we settled on the Hill Holliday ad agency of Boston, which had come up with a six-part series of print ads portraying innovatively shaped pin boxes, with copy that emphasized a CPA's breadth of experience and value. The pin boxes were shaped into images that appeared to rise in a manner that made them appear three dimensional. In one ad, for example, an image of ocean waves was superimposed by the tag line, "Even in the most treacherous business environment, the CPAs in your company can make sure you never get in over your head." Then the copy read:

Whether it's your CEO, CFO, or a department head, the CPAs in your company have the unique ability to help manage your company's resources. They can get beneath the numbers and help steer your company in the right direction. With strategic planning. Risk management. And even information systems and technology assessment. In short, they have the vision, expertise and pure brain power to keep your company out of deep water. And stay on course.

The other five ads had the following headlines:

- *A CPA can make the difference between being right on target or becoming one.* (An illustration of an arrow on a bull's eye.)
- *Assume the rocket is a company and someone incorrectly calibrated the guidance system. What was that you said about not needing a CPA?* (An illustration of a missile.)
- *After reviewing your retirement plans, your CPA strongly recommends mutual funds, short-term treasuries and the stone crabs at Emilio's in Palm Beach.* (An illustration of a sand crab.)
- *Some small companies try to design a retirement plan without a CPA. Some small companies have been out in the sun too long.* (An illustration of palm trees.)
- *The world wide web is here to stay. Would you rather be the spider or the fly?* (An illustration of a spider web.)

And all six ads concluded with:

You see numbers. We see opportunities.
THE CPA. NEVER UNDERESTIMATE THE VALUE.

The use of the pin boxes gave the ads a commonality and an organizational identification. The initial campaign began shortly after my term of service ended, during the first week of October 1995 and ran for six months, through March 1996. Each ad appeared in six national publications—Barron's, Forbes, Fortune, Inc., *The Wall Street Journal*, and *USA Today.* They were also provided to every state CPA society, many of which used their own financial resources to place them in local publications or local editions of national publications. A television commercial ran on Sunday morning business shows and various CNN programs.

We were under no illusions, of course, that any single image enhancement campaign of less than a year could dramatically transform businesspeople's attitudes about CPAs. But assuming we saw some progress, we were prepared to make a long-term commitment to the effort. Consequently, we were all gratified when in May 1996 a second Audits & Surveys study suggested that our initial $3 million ad campaign had been effective in several key areas.

Of the eleven business functions included in the research, perceptions concerning five had increased significantly. The post-ad campaign survey showed a fifteen percentage point increase in those business leaders who said that CPAs were the best profession to help with succession planning for family-owned businesses, and a ten percentage point increase in those saying we were the best choice to help businesses secure financing. Of the eight personal services, three showed significant improvement—estate, gift, and trust planning, family budgeting, and retirement planning. Perhaps most important, the number of respondents who agreed with the campaign's key tag-line, that CPAs "see more than numbers, they see opportunities," jumped by eleven percentage points.

The contribution of the state societies to the overall effectiveness of the campaign should not be underestimated. In fact, research revealed that the greater the media spending at the state level, the more positive the image of CPAs in that state became.

Given our modest initial budget, the ad campaign had proven unusually effective, and it was immediately renewed for another year, with a ten percent increase in the Institute's financial commitment. Subsequent studies confirmed that this national image enhancement campaign indeed had a positive impact on the attitudes of key decision makers who purchased CPA services.

MINORITY RECRUITMENT

In addition to reaching out to our external audiences during these years, we were also making a concerted effort to attract new recruits to the profession. Toward this end, the Institute sponsored a number of scholarships and other outreach programs to encourage the best and brightest students to consider accounting as a career choice. In 1993, for example, the Institute launched a new nationwide recruiting program centered around the slogan, "Accounting, Don't Just Learn a Business, Learn the Business World." Materials supporting the program were distributed to 50,000 high school and college educators.

Reporting on the initiative at our May 1993 Council meeting, Nita Dodson, chairman of the Accounting Careers Subcommittee of our Academic and Career Development Executive Committee, told us the accounting profession faced a shrinking college-age population and enrollment in accounting programs both in colleges and in business schools had declined. "The pipeline of new entrants into the profession must be filled, which is why we're starting our efforts at the high school level," she said.

The profession has also always prided itself on welcoming minority members. We like to believe we have been more progressive in this regard than most other professional organizations. In fact, on many occasions the Institute expressed its frustration that its aggressive attempts at achieving ethnic diversity have not been more successful.

The modern-day push to attract more minorities to the profession can be traced to a May 1969 AICPA resolution calling for full integration of the CPA profession "in fact as well as in ideal." In order to back up those words with action, the Institute implemented a variety of programs to encourage more minorities to study accounting and to use internships as a way of providing more employment opportunities for them during and after graduation.

Most of our efforts to increase minority participation were coordinated through the AICPA's Minority Recruitment and Equal Opportunity Committee. But since the early 1970s, the Institute has also been partnering with the AICPA Foundation in an array of programs to provide underrepresented minority students greater access to accounting education and to the profession. As a result, between 1980 and 1987 AICPA minority scholarships more than tripled, from just over $1 million to about $3.3 million.

At this same time, each of the Big Five firms were developing minority scholarships of their own. By the mid-1990s, the AICPA and individual firms had also begun to support other innovative programs, such as the St. Louis-based Inroads internship program. Each year Inroads finds employment in the accounting profession for about 600 black high school and college students.

Yet despite these efforts, I'm afraid we failed to attract as many minority accountants as we would have liked. From 1970 to 1990, the number of African-American CPAs climbed from 150 to 2500. In raw numbers this increase might seem impressive, but they continued to represent less than 1 percent of our membership. What's more, among all the college graduates hired by CPA firms in 1995, only 15 percent were nonwhite, and of that, 8 percent were Asian, 3 percent Hispanic, and 4 percent African-American.

Other professions weren't reporting much better results, of course. Only about three percent of medical doctors and four percent of lawyers in the United States are African-American. But according to a 1990 report in *USA Today*, only airline pilots and navigators had a lower representation of

African-Americans than did CPAs. In fact, African-American representation in non-minority owned CPA firms had actually declined, from 1.8 percent in 1976 to 1 percent in 1995. Even more disturbing was that in 1995, while 2.6 percent of all new hires were African-Americans, only 1 percent were partners. Hispanics constituted about 3 percent of hires, but also only 1 percent of partners.

Nevertheless, these less than stellar statistics by no means indicated that our efforts to recruit minorities were not worthwhile, or that they did not meet with some success. This was particularly true concerning the recruitment of minority accounting students, one of our top priorities in this area. In fact, African-American representation among accounting graduates has risen steadily since the AICPA began keeping statistics, from 5.6 percent in 1977, to 7.9 percent in 1995. And while the Big Five accounting firms continue to find it difficult to attract large numbers of minority CPAs, they have been much more successful in employing the efforts of minority accounting firms as subcontractors.

Another significant achievement was that in 1986, North Carolina A&T State University's School of Business and Economics in Greensboro became the nation's first historically black university to win accreditation for its undergraduate accounting program. African-American CPAs also slowly began to emerge into positions of responsibility and leadership—people like Bert Mitchell of Mitchell/Titus & Company, Dean Sybil Mobley of Florida A&M University, AICPA board of director Paula Cholmondeley, William Aiken of Peat Marwick, Clarence A. Davis of Oppenheim, Appel, Dixon & Company, and Watson Rice managing partner Tom Watson, to name just a few. Clarence has since become the chief financial officer of the AICPA.

As with other professions, the reasons for the paucity of African-Americans in the accounting profession have historically included racial bias and a lack of upward mobility, no strong affirmative action programs, and minimal awareness of career opportunities by minority youths. In our case, despite sincere efforts on the part of the largest firms to hire African-Americans and other minorities, we didn't do enough to motivate minority students to enter the accounting profession in the first place. This has been the single leading cause of our failure to attract more minority CPAs, and I know it is something on which the AICPA Minority Initiatives Committee continues to work. In order to be more successful, we just have to continue our current efforts, giving our full support to the committee, whose mission it is to "actively integrate minorities into the accounting profession ... and enhance their upward mobility." The Institute also continues to work with the National Association of Black Accountants, which represents about 5000 CPAs and about 60 minority-based accounting firms, to come up with innovative efforts that go beyond traditional scholarship programs.

In 1995, the Minority Initiative Committee launched an annual leadership workshop, at which AICPA scholarship students who were either college seniors or graduate students were invited to participate in workshops designed to develop leadership, presentation, communications, and team-building skills. Exposure to minority professionals and to other successful students is a key element of the program. The idea is that each successful minority CPA can inspire other students to join the profession. The state societies also developed programs during these years aimed at providing students with role models.

WOMEN CPAS

The accounting profession has had much greater success in attracting women CPAs. In fact, in 1980 one-third of all the accountants recruited by public accounting firms, and 37 percent of all accounting graduates, were women. By 1992, both those percentages had topped 50 percent, where they have remained.

These trends, of course, ran parallel to the overall integration of the U.S. work force, as by financial necessity the two-income family replaced the stereotype of the working dad and stay-at-home mom. Unfortunately, the accounting profession also mimicked society in that women continued to be significantly underrepresented when it came to leadership positions.

In 1995, while half of all graduating students and new hires were women, they made up only about 22 percent of the AICPA's membership. They were also underrepresented within top-level, managerial positions. In 1986, for example, an AICPA survey found that female managers occupied only 17 percent of all managerial positions in firms with 25 or more AICPA members. This was up from 14.3 percent in 1983 and only 4.7 percent in 1980, but still low. The number of partners told a similar story. Within the eight largest accounting firms, although the number of female partners more than doubled between 1983 and 1986 (from 69 to 157), they still represented only about 3 percent of the total.

We took numerous steps to try to rectify this situation. Once again we took our cue from the 1984 future issues report, which identified the upward mobility of women as one of the fourteen major issues facing the profession. The report made the following observations:

- Women had risen to various levels in accounting firms, but relatively few had become partners. This was the case despite the fact that the supply of qualified women nearly equaled the supply of qualified men.
- There was no question about women's technical ability, but stereotypes regarding other abilities raised the question whether they are at a competitive disadvantage to men in obtaining new audit clients and maintaining existing ones. In a kind of Catch-22, women's careers were also inhibited by the perception that men are generally more dedicated to a professional career than women.
- Professional women find advancement difficult due to society's pressure on them to be the center of the family unit. If a woman is to fulfill both professional and family responsibilities, she will need flexibility in her professional work—flexible hours, flexible workdays, and flexible work sites.
- The employee turnover rate within CPA firms is greater for women than for men, which is attributable at least in part to personal frustration.

As a direct result of the Future Issues Committee's report, the AICPA board of directors authorized the formation of a Special Committee on the Upward Mobility of Women. It was charged with "recommending strategies to strengthen the upward mobility of professional women who are employed not only in public accounting, but also in industry, government, and academia."

In its first report, issued in 1988, the committee identified a number of obstacles that were blocking the professional advancement of women accountants. The most obvious one, of course, was that in all professions, not just accounting, discriminatory attitudes toward women still existed in the workplace. More specifically, the committee described what it termed a "perception problem," where employers routinely denied the existence of any obstacles to the career advancement of female employees. This, the report suggested, left women feeling isolated. They typically did not discuss their problems with anyone for fear that management would give them an unfavorable performance evaluation. And while women, like men, were able to identify the personal traits necessary for advancement (technical competence, a willingness to work hard, good interpersonal and management skills, and leadership characteristics), many women were unable to identify the more subtle criteria necessary for advancement to top-level positions, such as having a mentor, visibility within the organization, and a successful self-image.

The committee offered eleven recommendations designed to help AICPA members and their employers eliminate these obstacles. Taken together, they laid out a plan to raise the profile of the female CPA. Within the AICPA itself, more women needed to be appointed to AICPA committees and boards, and on the Council. Steps also needed to be taken to consider the upward mobility of women when developing CPE courses, and to encourage state societies to appoint more women to leadership positions. The AICPA was directed to provide guidance for members and their employers on how to encourage the upward mobility of female professional staff. And on an ongoing basis, the Institute was urged to establish a process to monitor the report's recommendations.

The report also laid out a number of suggestions for improving women's chances for promotion, including having CPA firms encourage mentor systems so that new members of the profession would be able to discuss career-related issues with another, more experienced woman. It also urged female CPAs to improve their own career paths by becoming more active in client development activities and by promoting themselves within the profession.

During the years after the upward mobility of women report was issued, the committee worked hard to encourage employers to make the work environment more flexible. As if to formalize this focus, in the spring of 1992 the AICPA's governing Council changed the name of the committee to the Women and Family Issues Executive Committee. Not only did this give it permanent, more consequential status, but the name change reflected the Institute's intention to create a dual mission. It was the nineties, and the same lifestyle issues that were challenging women were now also confronting men. The new committee's mission, therefore, was not only to promote the upward mobility of women. It was also to help "men and women achieve a balance among personal, family and professional responsibilities."

One of the committee's first major initiatives was a detailed survey and analysis concerning the obstacles to the upward mobility of women CPAs. As reported in the October 1994 *Journal of Accountancy*, it debunked several misconceptions that had been holding back the careers of female accountants. Many managing partners, the committee found, assumed that turnover among women couldn't be affected by firm policy because it was a result of a woman's personal choice to make a change in her lifestyle. Many also thought that female accountants were less committed to their employment than were males. But neither of these beliefs held up under scrutiny.

These misperceptions were subtly destructive. Managing partners who subscribed to the notion that "after the baby, they're gone," admitted that as a result they might not invest as much in developing the career potential of female employees. In fact, the study found that 89 percent of female accountants return to their accounting jobs after maternity leave. But many firms were unprepared to manage those who wanted to amend or interrupt their careers until their children reached school age. The committee report noted a study by the Institute of Personnel Management at Ashridge Management College in Great Britain which indicated that the career path toward partnership in public accounting requires about ten years, and that stepping out of the cycle at any point was extremely detrimental to advancement. Creative, family-friendly policies, the committee report suggested, could help keep critical female employees from leaving.

Gradually, family-friendly policies that encouraged flexibility began to reduce employee turnover within the middle ranks of employment, and therefore increase the possibility that female CPAs would subsequently enter senior management. By the late 1980s, all but the smallest firms had initiated programs that built flexibility into career paths. Four-day work-weeks for mothers who wanted to stay on the partnership track, flextime, and formal maternity leave policies all helped increase overall participation by female accountants at every level of our profession.

Clearly, we had came a long way from the days when an editorial in the December 1923 *Journal of Accountancy* could declare that "women are not wanted as ... public accountants" because they would have difficulty serving "whenever and wherever called on to do so," or "working at night in places of difficulty and inconvenience."

In 1980, there was only one female member of the AICPA's board of directors. In 1995, there were five. In 1994, in the over-65 age category, more than 88 percent of AICPA members were men, but almost half under the age of 26 were women. In fact, as age declines, women were an increasing proportion of the membership in every age category. That bodes well for the future.

Women's partnership aspirations were also buoyed during these years by several Supreme Court cases. In 1984, in *Hishon v. King & Spalding*, the court ruled that partnership decisions fall under Title VII of the Civil Rights Act for firms with fifteen or more employees. This meant that discrimination with regard to promotion to partnership on the basis of race, color, religion, sex, or national origin was illegal. As a result, CPA firms have had to document their reasons for not admitting someone to partnership.

Then in 1985, in *Hopkins v. Price Waterhouse*, the court declared that a partnership may not inject stereotyped assumptions into its selection process. In this landmark case, Ms. Hopkins's work had been acknowledged by the firm as being "exceptionally good," but she had been criticized for being "overbearing" and "too assertive for a woman." The court ruled that Price Waterhouse had violated the law by evaluating her not as a prospective partner, but as a prospective female partner.

THE MOVE TO HARBORSIDE

As the chief administrator of a vast organization like the AICPA, much of the president's day-to-day routine is taken up by rather mundane matters. From a strictly financial point of view, one of the more important actions we took during these years was to move much of the Institute's operation out of midtown Manhattan.

The Institute had had its offices on three floors at 1211 Avenue of the Americas in New York City. By the mid 1980s, however, New York City real estate prices had skyrocketed and we had begun to look for alternatives. In May 1991, with the leadership and support of the late board chairman Gerald Polansky, we announced a relocation of 600 of our 800 employees to the Harborside Financial Center in Jersey City, New Jersey.

Not all our employees were overjoyed about the move, of course, since it meant the majority of our work force would now be working across the river from the Big Apple. Initially I was also concerned that the lack of opportunity for face-to-face meetings might create management problems. But a lot of the communications within and among our various divisions and committees was conducted by telephone anyway, so in the end the move really didn't involve as much of an organizational shift as I had feared.

Besides, the cost savings—an estimated $125 million over 20 years—were simply too good to pass up. Half the savings resulted from a reduction in rent, while most of the other half from the tax incentives New Jersey was giving us. As James Florio, the New Jersey governor at the time, said at the press conference announcing the move, "When we think accountants, we think bottom line."

Conclusion

Between 1980 and 1995, amid all the turmoil related to the litigation crisis and the powerful pressures to expand the responsibilities of CPAs, foundations were established by the profession for its future. That the AICPA was able to react to this whirlwind of activity with substantive change is a credit to the thousands of men and women who labored so tirelessly during these years to improve our profession.

Underpinning our efforts during these years was our focus on strategic planning, which evolved into the innovative work now being done by the AICPA's Vision Project. One of the many benefits we gained from our focus on future issues was an early emphasis on the use of technology, both within the Institute and among our members. This was formalized by the creation of a technology division. Today it would be difficult to imagine any CPA providing services without the daily use of technological innovations only dreamt of in 1980.

Another important success was the maturation process that occurred in the way we interacted with the federal government. With the intense scrutiny we had been under from the FTC and the U.S. Congress largely behind us, passage of the 1995 Private Securities Litigation Reform Act was the culmination of a decade-long effort to improve the effectiveness of our efforts in Washington. It represented a new political sophistication that continues to this day. On another front, an equally important part of our maturation process was, of course, implementation of the Anderson committee's recommendations.

Many challenges that seemed intractable during these years are still with us, but this is not necessarily a bad thing. The expectation gap—the continual process of considering our proper role in examining financial statements while educating the public to the limitations of an audit—is a never-ending struggle. Likewise, the manner in which we regulate ourselves is an ongoing responsibility that demands our constant vigil. It quite rightly is not an issue that can ever be "solved."

We could be doing a better job in some other areas, not the least of which is maintaining a greater diversity among our members. Between 1980 and 1995, the percentage of female accountants recruited by public accounting firms jumped from less than one-third to more than one-half. The profession had less success in attracting minorities, however, and much work in that regard still remains.

Just as businesses must constantly reinvent themselves in this fast-paced, technological world, so too must the accounting profession. That can be accomplished not only by fulfilling our responsibilities to the public by making certain that the information and services we provide remain relevant to the needs of investors, but also by doing all we can to help our clients prosper, all within the strict standards of professionalism expected of certified public accountants. So far, we have been able to do this without sacrificing the objectivity, integrity, and competence that the accounting profession has always held so dear. This, I believe, is the legacy that the years 1980 to 1995 bestowed on the profession.

As we look to the future, there is much uncertainty. What is the role of the auditor in an information-based economy in which historical financial statements seem to be less relevant? How can we serve the needs of the public for objective information, while at the same time fulfilling the desires of clients for professional advice and counsel, all within the constraints of independence? What will be the impact on the profession as firms of all sizes seek to remove advisory services from their organizational structure and take advantage of Internet commerce?

I am gratified to see the Institute and its leadership moving to meet the challenges of tomorrow. Helping members adapt to change is critical to their survival and to the survival of the AICPA, a truly great organization.

Appendix

AICPA CHAIRMEN, 1980–1995

1979 1980	William R. Gregory
1980–1981	William S. Kanaga
1981–1982	George D. Anderson
1982–1983	Rholan E. Larson
1983–1984	Bernard Z. Lee
1984–1985	Ray J. Groves
1985–1986	Herman J. Lowe
1986–1987	J. Michael Cook
1987–1988	A. Marvin Strait
1988–1989	Robert L. May
1989–1990	Charles Kaiser, Jr.
1990–1991	Thomas W. Rimerman
1991–1992	Gerald A. Polansky
1992–1993	Jake L. Netterville
1993–1994	Dominic A. Tarantino
1994–1995	Robert L. Israeloff

BOARD OF DIRECTORS, 1980–1995

John D. Abernathy	(1988–1990)
Brenda T. Acken	(1989–1993)
Jerrell A. Atkinson	(1994–1995)
Lowell A. Baker	(1988–1991)
Louis J. Barbich	(1992–1995)
Bernard Barnett	(1985–1986)

George L. Bernstein	(1983–1984, 1986–1987)
Robert L. Bunting	(1988–1990)
Philip B. Chenok	(1980–1995)
Paula H.J. Cholmondeley	(1984–1988)
James T. Clarke	(1993–1995)
Ronald S. Cohen	(1990–1995)
Diane S. Conant	(1993–1995)
J. Michael Cook	(1985–1987)
W. Thomas Cooper, Jr.	(1994–1995)
Sidney Davidson	(1986–1987)
Irvin F. Diamond	(1985–1986)
Leonard A. Dopkins	(1988–1991)
Dennis E. Eckart	(1992–1995)
Kathy G. Eddy	(1994–1995)
James Don Edwards	(1980–1982)
Merle S. Elliott	(1984–1988)
Robert K. Elliott	(1991–1992, 1994–1995)
Robert C. Ellyson	(1984–1987)
Ellen J. Feaver	(1992–1995)
Thomas M. Feeley	(1990–1994)
Barry B. Findley	(1983–1987)
John L. Fox	(1980–1982)
Barbara Hackman Franklin	(1980–1986)
Ray J. Groves	(1980–1985)
Bruce J. Harper	(1989–1993)
Gerald W. Hepp	(1983–1985)
Thomas L. Holton	(1983–1986)
Larry D. Horner	(1989–1991)
Kenneth J. Hull	(1991–1993)
Francis A. Humphries	(1988–1989)
Glenn Ingram, Jr.	(1983–1985)
Robert L. Israeloff	(1988–1990, 1993–1995)
Thomas G. Jordan	(1992–1995)
Charles Kaiser, Jr.	(1985–1989)
Charles E. Keller III	(1992–1995)
John H. Kennedy	(1993–1995)
Stuart Kessler	(1991–1994)
Olivia F. Kirtley	(1994–1995)
Paul Kolton	(1991–1995)
Raymond C. Lauver	(1980–1983)

Bernard Z. Lee	(1980–1984)
Ulysses J. LeGrange	(1986–1989)
Herbert J. Lerner	(1989–1993)
Alan B. Levenson	(1983–1990)
Herman J. Lowe	(1980–1986)
Andrew P. Marincovich	(1983–1984)
Robert D. May	(1980–1981, 1988–1991)
W. Ian A. McConnachie	(1983–1985)
Robert Mednick	(1986–1989, 1992–1994)
John R. Meinert	(1985–1986)
Robert A. Mellin	(1980–1982)
J. Curt Mingle	(1991–1995)
A. Tom Nelson	(1991–1995)
Jake L. Netterville	(1990–1994)
Paula C. O'Connor	(1993–1994)
Aulana L. Peters	(1991–1995)
John Phalen	(1990–1991)
Richard E. Piluso	(1988–1992)
Gerald A. Polansky	(1986–1993)
Thomas C. Pryor	(1980–1983)
William R. Raby	(1983–1984)
William C. Rescorla	(1980–1982)
Thomas W. Rimerman	(1986–1992)
Mahlon Rubin	(1984–1987)
Ralph S. Saul	(1985–1991)
Eric L. Schindler	(1992–1995)
Donald J. Schneeman	(1983–1994)
Joseph A. Silvoso	(1983–1986)
A.A. Sommer, Jr.	(1980–1982)
Charles G. Steele	(1984–1985)
A. Marvin Strait	(1983–1989)
Don J. Summa	(1983–1986)
Sandra A. Suran	(1989–1992)
Dominic A. Tarantino	(1988–1994)
John P. Thomas	(1988–1989)
Robert D. Thorne	(1983–1984)
James C. Treadway	(1994–1995)
Doyle Z. Williams	(1988–1991)
Kathryn D. Wriston	(1986–1992)
Arthur R. Wyatt	(1983–1984)

AICPA OFFICERS
1980–1995

Year	Vice Presidents	Treasurer
1979–80	Raymond C. Lauver Robert A. Liberty Richard D. Thorsen	Harry R. Mancher
1980–81	A. Marvin Strait Jon J. van Benten Arthur R. Wyatt	William B. Keast
1981–82	Sam I. Diamond, Jr. Arthur J. Dixon George E. Tornwall, Jr.	William B. Keast
1982–83	Thomas L. Holton Sybil C. Mobley Willis A. Smith	William B. Keast
1983–84	Barry B. Findley William L. Raby Robert D. Thorne	Don J. Summa
1984–85	Paula H.J. Cholmondeley Merle S. Elliott Charles G. Steele	Don J. Summa
1985–86	Bernard Barnett Irvin F. Diamond John R. Meinert	Don J. Summa
1986–87	George L. Bernstein Robert L. Bunting Sidney Davidson	Gerald A. Polansky

Year	Vice Presidents	Treasurer
1987–88	Lowell A. Baker Leonard A. Dopkins Doyle Z. Williams	Gerald A. Polansky
1988–89	Francis A. Humphries Richard E. Piluso Dominic Tarantino	Gerald A. Polansky
1989–90	Brenda T. Acken Bruce J. Harper Herbert J. Lerner	Richard E. Piluso
1990–91	Ronald Cohen Thomas M. Feeley Jake L. Netterville	Richard E. Piluso
1991–92	Kenneth J. Hull J. Curt Mingle A. Tom Nelson	Richard E. Piluso

AICPA GOLD MEDAL RECIPIENTS
1980–1995

1980	Wallace E. Olson		
1981	Herbert E. Miller	Walter J. Oliphant	
1982	Michael N. Chetkovich		
1983	Walter E. Hanson		
1984	William R. Gregory (posthumously)		
1985	Rholan E. Larson		
1986	Donald J. Kirk		
1987	Robert C. Ellyson	John R. Meinert	Robert A. Mellin
1988	George D. Andersen		

1989   William S. Kanaga
1990   Bernard Z. Lee
1991   Ray J. Groves   Thomas L. Holton
1992   A. Marvin Strait
1993   James Don Edwards
1994   Herman J. Lowe
1995   Philip B. Chenok

AICPA MEDAL OF HONOR RECIPIENTS
1980–1995

Year	Recipients
1980	Elmer Staats
1984	John J. McCloy
1986	Thomas C. Pryor
1990	William H. VanRessellaer (posthumously)
1994	Donald J. Schneeman

SELECTED BIBLIOGRAPHY

AICPA. 1985. *A Pathway to Excellence*. N.p.

AICPA. 1987. *Plan to Restructure Professional Standards*. New York: AICPA.

AICPA. 1996. *National Advertising Campaign: A Measurement of Its Effectiveness*. Audits and Surveys Worldwide, May.

AICPA. 2000. "Code of Professional Conduct." In *Professional Standards*, Vol. 1. New York: AICPA.

AICPA, Accountants' Legal Liability Special Committee. n.d. *Accountants' Liability—A Program for Legislative Reform*. New York: AICPA.

AICPA, Accounting Standards Executive Committee. 1994. *Disclosure of Certain Significant Risks and Uncertainties*. New York: AICPA.

AICPA, Accounting Standards Overload Special Committee. 1981. *Report*. New York: AICPA.

AICPA, Auditing Standards Division. 1989. *Implementing the Expectation Gap Auditing Standards*. New York: AICPA.

AICPA, Commission on Auditors' Responsibilities (Cohen Commission). 1978 *Report, Conclusions and Recommendations*. New York: AICPA.

AICPA, Financial Reporting Special Committee. 1994. *Improving Business Reporting—A Customer Focus : Meeting the Information Needs of Investors and Creditors* (Jenkins Committee Final Report). New York: AICPA.

AICPA, Future Issues Committee. 1984. *Major Issues for the CPA Profession and the AICPA*. New York: AICPA.

AICPA, Future Issues Committee. 1988. *Strategic Thrusts for the Future*. New York: AICPA.

AICPA, Implementation Monitoring Committee. 1989. *Report on the Implementation of the Recommendations of The Task Force on the Quality of Audits of Governmental Units*. New York: AICPA.

AICPA, National Commission on Fraudulent Financial Reporting (Treadway Commission). 1987. *Report*. N.p.

AICPA, Professional Examination Service. Research and Development Department. 1991. *Practice Analysis of Certified Public Accountants in Public Accounting*. New York: AICPA.

AICPA, Public Oversight Board, Advisory Panel on Auditor Independence. 1994. *Strengthening the Professionalism of the Independent Auditor* (Kirk Report). New York: AICPA.

AICPA, Public Oversight Board of the SEC Practice Section. 1993. *In the Public Interest: Issues Confronting the Accounting Profession.* Stamford, CT: Public Oversight Board.

AICPA, Solicitation Special Committee. 1981. *Report.* New York: AICPA.

AICPA, Special Committee on Assurance Services (Elliott Committee). *Final Report.* New York: AICPA, Inc., 1997.

AICPA. Special Committee on the Future of CPA Continuing Professional Education (Terry Sanford, Chairman). 1993. *Report.* New York: AICPA.

AICPA, Special Task Force on Accreditation. 1994. *Report to AICPA Council,* May.

AICPA, Strategic Planning Committee. 1991. *Strategic Thrusts for the Future,* 2nd edition. New York: AICPA.

AICPA, Strategic Planning Committee. 1993. *Report.* New York: AICPA.

AICPA, Strategic Planning Committee. 1994. *Report.* New York: AICPA.

AICPA, Strategic Planning Committee. 1995. *Report of the 1994–95 Strategic Planning Committee.* New York: AICPA.

AICPA, Upward Mobility of Women Special Committee. 1988. *Report to the AICPA Board of Directors, March 1988.* New York: AICPA.

AICPA, Women and Family Issues Executive Committee. 1994. *Survey on Women's Status and Work/Family Issues in Public Accounting.* New York: AICPA.

AICPA and National Association of State Boards of Accountancy. *Uniform Accountancy Act, 1992 and 1994.* New York: AICPA.

American Accounting Association, Committee on the Future Structure, Content, and Scope of Accounting Education. 1986. "Future Accounting Education: Preparing for the Expanding Profession" (Bedford Report). *Issues in Accounting* 1(1, Spring).

Arthur Andersen & Co. 1992. *The Liability Crisis in the United States: Impact on the Accounting Profession.* Arthur Andersen & Co.

Audits and Surveys, Worldwide, Inc. 1995. *CPA Image Tracking Research: Top-Line Findings from the Benchmark Wave,* November.

Audits and Surveys Worldwide, Inc. 1998. *Measuring the Effectiveness of the 1997–98 CPA Image Enhancement Campaign: A Final Report on the Research Findings,* June.

Carey, John L. 1969–1970. *The Rise of the Accounting Profession.* 2 vols. New York: AICPA.

Chapin, Donald H., Robert W. Gramling, and John D. Dingell. 1996. *The Accounting Profession—Major Issues: Progress and Concerns.* 2 vols. Washington, DC: U.S. General Accounting Office, Accounting and Information Management Division.

Commission on Professional Accounting Education (Wayne Albers, Chairman). 1983. *Implementation of a Post-Baccalaureate Education for the CPA Profession, for the CPA Profession.*

Committee of Sponsoring Organizations of the Treadway Commission. 1994. *Internal Control—Integrated Framework.* 2 vols. New York: AICPA.

Coopers & Lybrand. 1994. *Audit Committee Guide.* Coopers & Lybrand.

Derieux, Samuel A. 1980. *Report of the Special Committee on Small and Medium Sized Firms.* New York: AICPA.

Hart, Peter. Study commissioned by the National Accountant's Coalition, established in 1993 by the Big Six to investigate the liability crisis for themselves.

Kulberg, Duane R. 1989. *Perspectives on Education: Capabilities for Success in the Accounting Profession.* (Views of chief executives of the eight largest public accounting firms on their position on education for the accounting profession .)

Knutson, Peter H. 1993. *Financial Reporting in the 1990s and Beyond.* Charlottesville, VA: Association for Investment Management and Research.

Louis Harris and Associates. 1986. *A Survey of Perceptions, Knowledge, and Attitudes Toward CPAs and the Accounting Profession.* New York: AICPA.

Louis Harris and Associates. 1988. *Study on Allowing Certified Public Accountants to Accept Payments in the Form of Commissions and Contingent Fees,* July.

Magill, Harry T., Gary John Previts, and Thomas R. Robinson. 1998. *The CPA Profession.* Upper Saddle River, NJ: Prentice Hall.

McMullen, Dorothy A., and K. Raghunandan. 1996. "Enhancing Audit Committee Effectiveness." *Journal of Accountancy* 82(August): 79–81.

McMullen, Dorothy A., K. Raghunandan, and D.V. Rama. 1996. "Internal Control Reports and Financial Reporting." *Accounting Horizons* 10(December): 67–75.

Miller, Richard I., and Michael R. Young. 1987. "Financial Reporting and Risk Management in the 21st Century." *Fordham Law Review* 65(April): 1987–2064.

New York State Society of CPAs, Special Advisory Task Force to Study Blacks in the CPA Profession. 1990. *Report on the Status of Blacks in the CPA Profession.* New York: Author.

Olson, Wallace E. 1982. *The Accounting Profession: Years of Trial: 1969–1980.* New York: AICPA.

Previts, Gary John, ed. 1991. *Financial Reporting and Standard Setting.* New York: AICPA.

Price Waterhouse. 1985. *Challenge and Opportunity for the Accounting Profession: Strengthening the Public's Confidence: The Price Waterhouse Proposals.* New York: Price Waterhouse.

Price Waterhouse. 1993. *Improving Audit Committee Performance: What Works Best.* Altamonte Springs, FL: Institute of Internal Auditors Research Foundation.

Rimmerman, Thomas W. 1990. "The Changing Significance of Financial Statements—The Relative Disparity Between Content and Needs: 1988 Future Issues Committee Report." *Journal of Accountancy* 169(April): 79, 82–83.

Stevens, Mark. 1981. *The Big Eight.* New York: Macmillan.

United States v. Arthur Young & Co., 1984.

U.S. Congress. *Racketeer Influenced and Corrupt Organizations Act (RICO).*

U.S. General Accounting Office. 1991. *Audit Committees: Legislation Needed to Strengthen Bank Oversight.* Washington, DC: U.S. GAO.

U.S. General Accounting Office. 1989. *CPA Audit Quality: Status of Actions Taken to Improve Auditing and Financial Reporting of Public Companies.* Report to the Chairman, Oversight and Investigations Subcommittee, Committee on Energy and Commerce, House of Representatives. Washington, DC.

U.S. House of Representatives, Committee on Government Operations. *Substandard CPA Audits of Federal Financial Assistance Funds: The Public Accounting Profession is Failing the Taxpayers* ("Brooks Report"). 1986. H.R. 99–970. 99th Cong., 2d Sess., 1986. Washington, DC : U.S. Government Printing Office.

U.S. House of Representatives. 1986. *Financial Fraud Detection and Disclosure Act* (H.R. 5439).

U.S. Senate. *Private Securities Litigation Reform Act.* 1994. S.1976. 103rd Cong., 2d sess.

Williams, Doyle Z. 1994. *Accounting Education: A Statistical Survey 1992–1993.* New York: AICPA.

Young, Michael R. 1996. "The Liability of Corporate Officials to Their Outside Auditor for Financial Statement Fraud," 64 *Fordham Law Review* (April): 2155–2184.

Young, Stephen J. 1995. *The Changing Profile of the AICPA: Demographics of a Maturing Profession.* In *Research in Accounting Regulation,* vol. 9, ed. Gard John Previts. Greenwich, CT: JAI Press.

Index

Printed in the United States
83111LV00003B/46/A

9 780762 306725